ANCIENT ALIENS ON MARS II

MIKE BARA

Adventures Unlimited Press

Ancient Aliens on Mars II

by Mike Bara

ISBN 13: 978-1-939149-31-2

Published by:
Adventures Unlimited Press
One Adventure Place
Kempton, Illinois 60946 USA
auphq@frontiernet.net

www.AdventuresUnlimitedPress.com

ANCIENT
ALIENS
ON
MARS
II

Adventures Unlimited Press

Acknowledgements

I would like to acknowledge the following souls for their roles in my life during the writing of this book: Dave Bara, Denise Zak, Robert Quicksilver, Serena Wright Taylor, Tom Danheiser, George Noory, David Hatcher Childress, Jason Martell, the gang at RAW TV, everybody on the Ancient Aliens staff, TabbyKat, Sherri (thanks for the sweater), my co-stars on Uncovering Aliens, my ex-Facebook wife Sara Vasquez, Stephanie Mann and Sheila Knies, and of course Aurora, Miss Fluffy-Muffy, Seabass and Barkley Bear.

TABLE OF CONTENTS

Dedication

This book is dedicated to Denise Zak, the wonderful, supportive sister I wish I'd had growing up. May God's light always shine upon you. I love you.

See Mike Bara at:

MikeBara.Blogspot.com

Author's Notes

The images in this book may be found in full high resolution on web site at http://mikebara.blogspot.com/. Simply click on the "Photos" tab and follow the links to my Picasa Web Album.

Every effort has been made in this volume to give proper credit to the discoverer of any objects or anomalies not first noted by the author. If you feel that an error has been made in assigning proper credit for a discovery, please contact my publisher Adventures Unlimited Press and every effort will be made to correct the record in subsequent editions of this book.

Mike Bara 5/22/2014

Acronyms used in this Book

IR — Infrared

MRO — Mars Reconnaissance Orbiter

MGS — Mars Global Surveyor

THEMIS — Thermal Emission Imaging Spectrometer

RAT — Rock Abrasion Tool

ESA — European Space Agency

HRSC — High Resolution Stereo Camera

MOC — Mars Orbiter Camera

HiRise — High Resolution Imaging Science Experiment

MPP — Meters-Per-Pixel

MSSS — Malin Space Science Systems

SPSR — Society for Planetary SETI Research

Introduction

When I first began my previous book, *Ancient Aliens on Mars*, it was my intention to make it the seminal, complete "go to" document of all of the weird, anomalous and quite obviously artificial structures and monuments there. I was surprised to discover about half way through that I wouldn't even get close. Just telling the full story of the ruins at Cydonia and the Face on Mars would take up far more space than I could possibly put in one book, especially one with a tight word count and a strict release deadline. Hence this book was born.

That is not to say we won't continue and finish with the Face on Mars in this volume. We will. But I am also anxious to look beyond Cydonia to the many other mysteries of Mars, to questions of life, fossils and ruins in other regions. We will do all that on this journey. But we will also go beyond that and look at some other myths and legends of the Red planet as well. There has never been a comprehensive look at the mysterious case of the Russian probe Phobos 2, for instance. It simply disappeared, as so many Mars probes do, while taking images of both the surface of the planet (which as we established in book one is actually a moon) and what appeared to be "something" in orbit with the spacecraft. That "something" has led to decades of speculation about there being some kind of automated defense system—something like the old abandoned machines of the Krell race in the classic Sci-Fi movie *Forbidden Planet*—protecting Mars from unwanted interlopers.

We will also look at the many strange findings of NASA's more recent rovers, Spirit, Opportunity and Curiosity. We will turn a careful eye towards images of the "rats," "lizard's" and "figurines" on Mars that prompted SETI's Seth "don't call me SET" Shostak to write his silly and vacuous *Huffington Post* blog "Martian Archaeology—Not" when many of these objects were first pointed out. We will dig much deeper than the superficial dismissals of the NASA apologists, and get to the heart of these subjects no matter where the data may lead us. That's what real inquiry is all about. Simply asking questions, and not accepting simple, t-shirt slogan answers like "extraordinary claims require extraordinary proof."

But I will warn you now that as intriguing as many of these speculative scenarios are, most simply will not hold up under the weight of the evidence. I won't tip you off here, but there are some enigma's that were more fun to contemplate before we looked too close.

That is not to say we won't find more genuine mysteries in these pages—we will. It's just that some long held beliefs about Mars and what it holds will turn out to less than their discoverers originally thought. But there is also far more high strangeness about Mars that has never been properly acknowledged, the work of numerous independent and dedicated researchers who have scoured the web and looked at countless images until they found something that made them exclaim, just as Toby Owen did at JPL in 1976 when he first spotted the Face on Mars, "Hey, look at this!"

Trust me, there will be many more "Hey look at this!" moments in this book. But as I said, there will also be old mysteries and myths that will be put to rest once and for all. Our task on this journey, yours and mine, must be to discern the difference between the two so that we can move toward the only real value and intent of these books—the truth.

Because the truth is something that is in short supply today. From the crass intellectual dishonesty of "Martian Archaeology, not" to the outright, bald-faced mendacity of "If you like your plan, you can keep it," truth has become a rare commodity. Even in the hallowed halls of science, which is supposed to be ultimately a search for truth, many, many lies are told and retold and inculcated into the minds of mush-brained young students in our universities. This in turn gives rise to a complete lack of true intellectual curiosity, and a new and sadly inept generation of "scientists" who have been taught *what* to think, not *how* to think. Fueled at every turn by the ridicule of the debunking elite like James Oberg and Phil Plait and kept in line by the NASA controlled funding streams, we have reached a point where the reality of an Ancient Alien presence on Mars cannot be rationally denied, yet at every turn it is. The only way to combat this is to simply stand in your own truth, to know what you know and resist the pressure to conform.

We will never find that truth though in the false hope of the silly "disclosure movement." Even if those that push this hopeless agenda fail to understand its inherent fallacy. Public disclosure of the type that they demand not only puts public servants *above* the people—which they most certainly are not, it naively assumes that what is "disclosed" by these inveterate liars will be the actual truth.

It will not.

Even if some sort of official government "disclosure" takes place, which I do not believe will ever happen, it will at best be some sanitized version of the truth, and one designed to do nothing more than to secure those making the disclosure in their current positions of power.

So no, this book is not about "disclosure." It is about truth. It is about the kind of truth that you can only know in your heart and in your mind. And it is the kind of truth that doesn't need the approval of your distracted, cell phoned zombie apocalypse friends or an out of control and inherently untruthful government.

And it is about one other thing. It is about giving you the ammo you need to point out to the dismissive friends and family members, the scientific conformists and the elitist snobs you've had to put up with that you aren't the weird ones for believing in all this stuff, *they* are. They are the ones that are so focused on their games and their phony politics and their shrinking job security that the truth has simply stood up and walked right by them without even saying hello.

With all that in mind, we will start off this journey where we left off the last one. On the enigmatic plains of Cydonia…

Chapter 1
Monoliths and Mysteries

When we left off in *Ancient Aliens on Mars*, we had covered quite a bit of the data and debate around Cydonia and the Face on Mars. We had looked at the early *Viking* data, poured through the skullduggery surrounding the infamous "Catbox" image and dug down deep into the *Mars Global Surveyor* (MGS) images of the Face. Beyond that, we looked a bit into the European Space Agency's (ESA) *Mars Express* imagery and examined the exquisite detail of the *Mars Reconnaissance Orbiter* (MRO)images of the Face on Mars. But where we hadn't spent quite as much time was

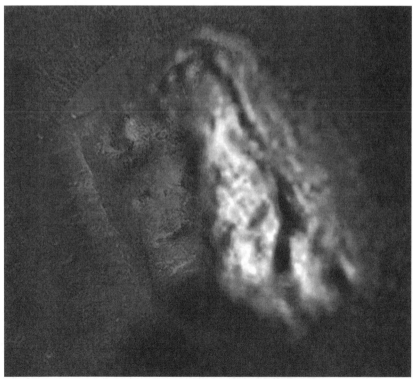

Combined THEMIS pre-dawn thermal infrared and MGS visual images of the Face on Mars.

on the *2001 Mars Odyssey* mission and its suite of instruments that looked at Cydonia and the Face in a whole new way; specifically, with a thermal infrared imager named THEMIS.

THEMIS, which stands for Thermal Emission Imaging System, was the primary scientific instrument on the *Mars Odyssey* spacecraft, and contained a medium resolution visual imager as well as a thermal infrared imager that operated in nine different infrared wavelengths of the electromagnetic spectrum.

As these things go, the medium-res visual camera was quite a good scientific instrument, certainly better than the imager on the *Viking* probes but not quite as good as the "Malin camera" mounted on the MGS or the later Hi-RISE camera on the MRO. Whereas MGS could get theoretical resolutions below five meters-per-pixel under ideal conditions, images taken by the visual band camera on *Mars Odyssey* fell into the 14-20 MPP range. That range is good enough to resolve a semi tractor-trailer vehicle, and more than sufficient to identify key features of the Face as small as the disputed "pupil" in the right side eye socket, which at about 550 feet across is as big as baseball stadium. Below five MPP, objects like large trucks and cars become resolvable, while it takes resolutions below .5 meters-per-pixel to distinguish an individual human being.

The first *Mars Odyssey* visual image of the Face on Mars

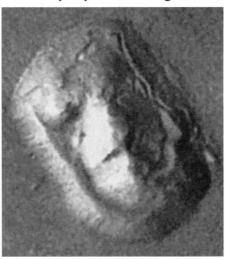

Mars Odyssey 2001 visual image of the Face on Mars

turned out to be about 18 meters-per-pixel, substantially better than the *Viking*-era imagery.[1] Had such an image been available in the 1990s, it would have more than satisfied the demands of the independent Mars research community. It would have also deepened the debate about artificiality.

What the image showed was that the earlier (and supposedly "better") *Mars Global Surveyor* images of the Face had been grossly distorted. Because it was taken more directly overhead than the MGS images, the *Mars Odyssey* image showed the size, shape and feature arrangement of the Face far more accurately and in better detail, without the distortion caused by the orthographic rectification processes NASA used. It showed, among other things, that the first level "platform" upon which the Face was constructed was almost uniformly symmetrical, in the range of 95%. It had two parallel edges of almost exactly the same length which ran for hundreds of meters on each side, and it was capped by two equally perfect arcs of nearly the same radius at the top and bottom. These features alone would have identified the Face as most probably artificial and worthy of an archeological survey had it been discovered on Earth.

But the image also showed other features in better relief. The Face was not nearly as wide as it appeared in the *Global Surveyor* images. The two eye sockets, which had appeared to be in different locations on the MGS images, were shown to be in direct alignment

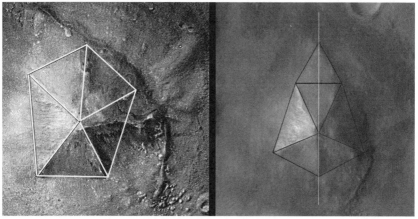

Bilateral symmetry of the D&M pyramid as reconstructed by Torun (left) and as seen by *Mars Odyssey*'s THEMIS instrument.

with each other, as if someone had built them that way. There was also the hint of a reappearance of the "nostrils" in the nose first identified on the "Catbox" MGS image.

The wide angle THEMIS image also gave us very clear views of the full context of the entire Cydonia complex, and revealed previously unseen details in many of those objects as well. Chief among these was the D&M pyramid.

The original *Viking*-era data had revealed, under examination by Defense Mapping Agency expert Erol Torun that the D&M was once bilaterally symmetrical about an axis formed by an edge that pointed directly at the Face on Mars. What the new, better resolution THEMIS image showed was that not only was *that* symmetry confirmed, but the D&M was actually formed (or more probably built) atop a huge circular rise or platform. It was also shown to be a much more complex and interesting geometric object than had been previously even considered. The signature of this was the discovery in the superior *Mars Odyssey* image that there was also a *2nd line of bilateral symmetry,* formed by a triangular "tail" at the base of one side of the D&M and the intersection of two matching faces that formed a 60 degree angle. This second line of symmetry is so decisively obvious as a signature of artificiality that any question of its natural formation must be regarded as far beyond the improbable and well into the preposterous. In simple terms, mountains, especially isolated ones, simply never have even *one* axis of bi-lateral symmetry, much less two. A second axis of bilateral symmetry is as certain a signature as one can have as proof of artificiality. To even argue the point would pretty much thoroughly discredit you as an unbiased observer.

So before we'd even gotten to the good stuff—the thermal infrared images—we'd already had a major boost to the artificiality hypothesis just with the visual images. But the infrared camera promised even more.

Because it operated in nine bands of the infrared spectrum, THEMIS could theoretically discover a great number of things that mere visual images could never reveal. It could not only determine differences in mineral composition and distribution in a given image, it could also measure the relative heat-emitting characteristics of those minerals. This could aid in determining not just that there *were* differences in the material compositions of various objects themselves, but even possibly *what* each of these different materials were made up of. Different rocks, metals and clays have different heat dissipation cycles, so if properly calibrated for the Martian environment, THEMIS held the potential of telling us exactly what the Face and other controversial objects on the surface were actually made of.

Infrared images could also provide a limited below ground imaging capability. Almost like ground penetrating radar, THEMIS could make it possible to "see" below the Martian surface, provided that whatever was beneath the surface was emitting a significant and

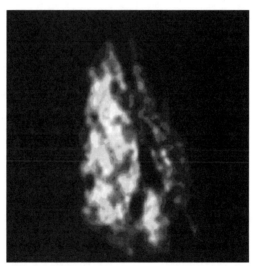

2003 THEMIS infrared image of the right, or eastern side of the Face on Mars showing highly reflective geometric panels making up the object.

measureable degree of heat.

These two capabilities made it an ideal instrument to study the Face on Mars and other unusual features on the Martian surface. That came in handy when the first pre-dawn THEMIS image of the Face was released in July 2003. As we covered in chapter eight of *Ancient Aliens on Mars*, that image revealed incredibly reflective geometric panels on the sunward side of the Face, aligned with the parallel geometry of the Face itself rather than the image scan direction. Details of these panels were faintly revealed in close-up study of MRO images of the Face, but were obscured by a thin layer

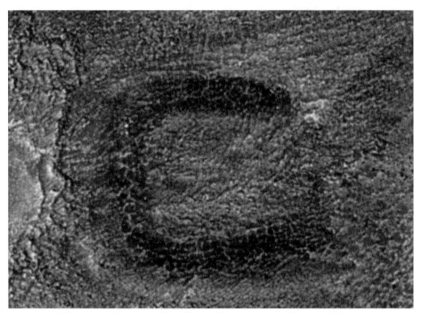

Close-up of structural collapse feature on the Face on Mars.

of dust which the THEMIS infrared camera was able to cut through.

Obviously, such a capable instrument had great potential to settle arguments about the artificiality of not just Cydonia, but other regions on the surface of Mars which were in dispute. All we needed, we thought, was to get nighttime thermal infrared images of Cydonia to decide the issue. Unfortunately, with the exception of the pre-dawn image, this data was not easy to come by. So we waited.

Mixed up in all this, as political pressure was applied to Arizona State University and NASA to take infrared images of Cydonia, was an enduring mystery of a previous attempt to take infrared images of Mars. In the late 1980s, the old Soviet Union had sent two probes, named *Phobos I* and *Phobos II,* to Mars to study the surface and atmospheric properties of the planet and the composition of one of its two moons, Phobos.

Phobos I failed along the way, but *Phobos II* made it all the way to Mars and operated nornally for period of several weeks. Its disappearance has become the stuff of UFO lore, but in the process the spacecraft made numerous valuable observations of both Mars and its "moon" Phobos. One of the most curious was that Phobos' density was found to be extremely unusual. According to a paper published in the October 19, 1989 issue of the scientific journal *Nature*, Phobos had a bizarre density of 1.95 grams per cubic centimeter, meaning it was almost *one third hollow.* Since both Martian "moons" were considered captured asteroids, this finding was mind-boggling. Since asteroids are, in the prevailing models, "accreted" clumps of dust and rock (see my book *The Choice*), there is virtually no way that a supposedly solid object like Phobos can be "hollowed out" in this manner naturally. That left a really big question—just who would have hollowed it out, and why?

The situation got even more intriguing when *Phobos II* took its first images of the moon Phobos, and discovered that it was even stranger than had been suspected. The close-up images revealed

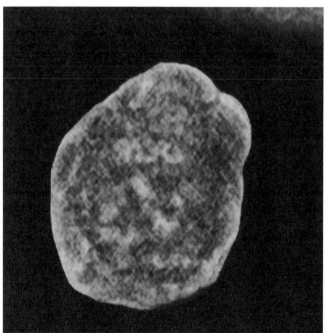

Phobos II image of Mars' moon Phobos. Note geometric, right-angled pattern running all across the body of the object.

19

-a bizarre, repetitive and distinctly geometric track pattern on the surface of Phobos, almost as if someone had been riding mega-sized golf carts back and forth across the surface at right angles to each other. This crisscross pattern was explained as orbiting boulders, slowly pulled down to the surface of Phobos by the moon's meager gravitational field, landing and running across the surface and leaving a trail.

Forgetting for a moment the fact that an asteroid with multiple orbiting satellites is a key prediction of Dr. Tom Van Flandern's Exploded Planet Hypothesis (see *Ancient Aliens on Mars* and *The Choice*), and is therefore in conflict with whole "accretion" theory of asteroid formation, this "explanation" failed to address the key

ESA *Mars Express* image of Phobos.

observation. Why would *all* of these boulder tracks be at right angles (90°) to each other?

The short and simple answer is they wouldn't. Not if they were formed naturally. So that left only one possible explanation: they weren't natural at all.

The whole issue was accelerated in March 2010, when the European Space Agency's *Mars Express* made a close fly-by of Phobos and took images and other measurements with the spacecraft's High Resolution Stereo Camera (HRSC). Blogger Richard C. Hoagland (my co-author on *Dark Mission*) quickly confirmed the "rectilinear" pattern seen in the earlier *Phobos II* mission images. His conclusion was that what the crosshatch pattern was showing was the underlying, exposed mesh-like grid pattern of Phobos' exterior, *artificial* shell.

This impression was reinforced by some later scientific papers, published in the journal *Geophysical Research Letters*, which supported the earlier Russian probes' conclusion that Phobos was, inexplicably, hollow…

Using two different instruments to measure both the mass and density of Phobos, *Mars Express* team scientists concluded (and admitted) that not only was Phobos one-third hollow, as the

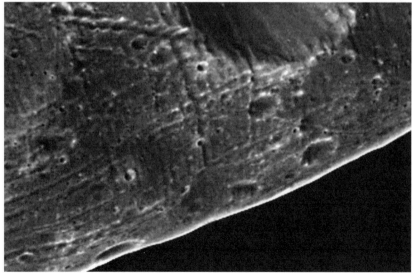

Close-up view of Phobos showing the crosshatching pattern in greater detail (ESA).

Russian probe *Phobos II* had also concluded, but also that it was most probably *not* a natural object:

> *New values for the gravitational parameter and density of Phobos provide meaningful new constraints on the corresponding range of the body's porosity (30% ± 5%), [and] provide a basis for improved interpretation of the internal structure. We conclude that the interior of Phobos likely **contains large voids**. When applied to various hypotheses bearing on the origin of Phobos, **these results are inconsistent with the proposition that Phobos is a captured asteroid.***

Allow me to translate for you: If Phobos is *not* a captured asteroid, which is the only natural explanation for its presence in orbit around Mars, then it by definition *must* be an *artificial* satellite. Although mainstream science internet sites have sought to downplay these results, claiming for instance that "most" voids in the interior of Phobos are small (two to three feet across), they gloss over the fact that there are several such voids well over 200 feet across. The ballroom-sized "voids" in the interior of Phobos can be only one thing: rooms. Call them condos, apartments, whatever you want. But they are rooms inside the structure of what appears to be an artificial body, placed in an equatorial orbit around Mars.

The whole issue of Phobos' artificiality was made even more confusing by the discovery by two independent researchers, Efrain Palermo and Lan Fleming, of what they called the Monolith on Phobos. The original, blurry *Mars Global Surveyor* image looked like a large boulder casting a long shadow on the surface of Phobos near the crater Stickney. Evoking memories of the tall, thin black object from Stanley Kubrick's film *2001: A Space Odyssey*, the Monolith quickly caught fire on the internet and was even mentioned as potentially artificial by NASA astronaut Buzz Aldrin in a famous 2009 C-SPAN interview. For the most part, it was dismissed as an optical illusion along the lines of the "Blair Cuspids" on the Moon (see *Ancient Aliens on the Moon*). That is, until something interesting happened. The much better Hi-RISE camera on the *Mars*

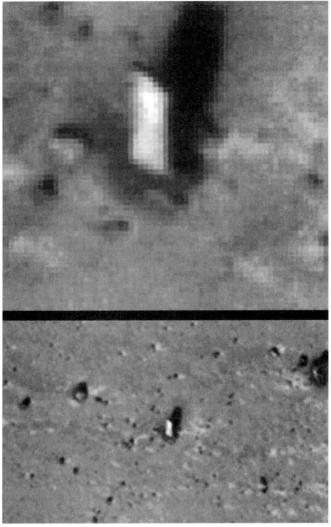

MRO image of the "Monolith" on Phobos

Reconnaissance Orbiter was pointed at it.

When that image was returned, the "Monolith" turned out to look, well, exactly like the Monolith from *2001*. It has a distinct, rectangular slab-like shape that is decidedly unnatural in appearance. Its presence jutting upright out of the sands of Phobos bore more than an eerie resemblance to the alien structure from *2001*; it was a dead ringer for it. A proposed Canadian mission to Phobos named PRIME has suggested the Monolith site as a possible landing site, but so far the mission has not been funded or had a launch date set.

But all of this, believe it or not, isn't even the most interesting controversy that *Phobos II* started with the data it returned. That happened when the spacecraft was rotated to point its thermal infrared imager at Mars itself.

Phobos II carried an infrared spectrometer, a device not too different from the infrared thermal imager today on *Mars Odyssey*. While it lacked the resolution of *Odyssey's* far better THEMIS camera, the infrared device on *Phobos II* still gave the Russian scientists the capability to discern buried objects just below the surface of the planet (covered with sand or dust) —via their relative rates of cooling.

And guess what?

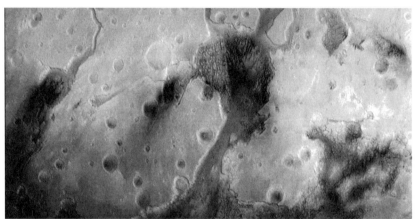

Phobos II thermal infrared image of what appear to be buried ruins in Hydroates Chaos on Mars.

Several of the thermal infrared images returned by the spacecraft showed a strange, geometrically arranged series of wedge-shaped structures just beneath the Martian sands. Located in a region known as Hydroate Chaos, the area bore a strong resemblance to buried archaeological ruins discovered on Earth.

While this area may just look like a jumbled mish-mash of wedges and blocks to some of you, it's important to keep in mind Carl Sagan's axiom about the search for life on Earth by an alien species. As he put in his book *Cosmos*: "The first indication of intelligent life on Earth lies in the geometric regularity of its constructions..."

What Sagan was expressing is that intelligence is defined by

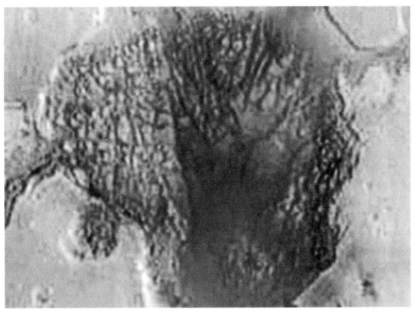
Close-up.

specific geometric markers when viewed from orbit. The outlines of its fortresses, the street layouts of its cities, the rectangular patterns of its farms and orchards. Everywhere you look on Earth, this signature of intelligent manipulation of the biosphere is obvious. Why shouldn't it be so on Mars as well?

Almost immediately after the return of this image from *Phobos II* in 1989, a controversy sprang up about it. The Russians immediately held a press conference in the wake of the image being received, and declared the area to be an artificial cityscape.

Video capture from press briefing at the Soviet Space Research Institute in 1989.

25

Speaking at a televised press briefing at the Soviet Space Research Institute, the Russian scientists told the assembled press unequivocally that *Phobos II* had returned an image of a buried city the size of Los Angeles on Mars. They showed the assembled international press the enhanced image mosaics from *Phobos II* and pointed out the anomalous, geometric features. They left no doubt about their interpretation of the data: they declared it to be a buried Martian city. Amazingly, the press briefing went virtually uncovered in the United States.

But it did get some coverage in Europe, and caught the eye of a British scientist, Dr. John Becklake. Becklake is the author of over 20 scientific papers on rocketry and spaceflight as well as three books and is a member of the International Academy of Astronautics. At the time he heard about the Soviet press briefing, he was the director of the London Science Museum and he immediately made arrangements with the Soviets to bring their scientists to the U.K. to repeat their press briefing. In preparation, Becklake was given a set of 35 black and white high-resolution images taken by *Phobos II* by the Soviet scientists. With this help, Becklake created a museum exhibit based on the images the Russians sent him. At the last minute, the Soviets got cold feet and refused to allow their scientists to fly to London, so Becklake went ahead with the briefing on his own. The press conference was covered by only one British TV channel, the independent Channel 4 station in the U.K.

British Channel 4 broadcast video on the *Phobos II* findings.

In the Channel 4 report, Becklake showed the images the Russian scientists had given him and made several strong statements about what, in his opinion, they revealed. "The city-like pattern is 60 kilometers wide and could easily be mistaken for an aerial view of Los Angeles. We have some very, very thin lines on the surface of Mars in the infrared, which means it's heat, and it's not visible, it's heat. These have a width of 3-4 kilometers wide. As for the question of what it is, I don't know, and the Russians aren't telling us."

To illustrate his point, Channel 4 broadcast a blow-up of the area in question. The strange, right angle patterns fully supported

Close-up of "Los Angeles" cityscape as seen in British Channel 4 report in 1989.

Becklake's conclusions about the region. The program went on to show excerpts from the Soviet Space Research Institute press briefing in which the anomalies were discussed.

So there's no question that the findings were considered very strange, if not a smoking gun, by credible members of the scientific community back in 1989. A few years ago, in an effort to confirm the work done by the Soviets over 20 years before, I was able to obtain copies of all the *Phobos II* images through my co-author on *Dark Mission*, Richard Hoagland, who got them from a Russian source. Although the camera on *Phobos II* was primitive compared to the THEMIS instrument, the data is fairly easy to work with and I was able to do my own image processing on it. In doing so, I was able to verify and confirm the earlier work of the Soviet research team.

My version of the image in question actually brings out more detail, thanks to the superior software tools available today. There is no question that the right angle pattern exists, and that it does look exactly like a vast cityscape.

Rectilinear patterns like this are almost universally recognized

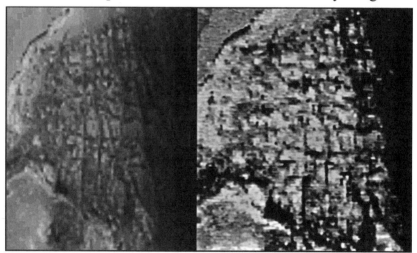

Phobos II image of the "City" in thermal infrared. Processed by the author (left) and as shown on Channel 4 (right).

as buried archeological sites—when they are spotted on Earth. Mars. apparently, is a different story...

Despite the scientific credentials of all the European scientific heavyweights involved in this story, it was once again virtually ignored in the United States, just as the original Soviet press briefing had been. It just wasn't reported. What *was* reported, in various fringe media and in UFO forums, was an event that happened almost immediately after *Phobos II* took the controversial image in question.

It disappeared.

On March 28, 1989, Tass, the official Soviet news agency released a statement regarding the disposition of *Phobos II*:

> *Phobos II* failed to communicate with Earth as scheduled after completing an operation yesterday around the Martian moon Phobos. Scientists at mission control have been unable to establish stable radio contact.

A day later, a top official of the Soviet Space Agency Glavkosmos (now the Russian Federal Space Agency) was quoted as saying "*Phobos II* is 99% lost for good."

Although the loss of the spacecraft was eventually ascribed to a computer failure, just a few days later rumors began that *Phobos II* had been lost due to a UFO encounter. On March 31, 1989, headlines issued by the Moscow correspondents of the European News Agency (EFE) stated: "*Phobos II* Captured Strange Photos of Mars Before Losing Contact With Its Base." The news article that followed gave even more sensational details. Quoting from the Soviet TV newsmagazine *Vremya,* it said:

> Vremya revealed yesterday that the space probe *Phobos II*, which was orbiting above Mars when Soviet scientists lost contact with it on Monday, had photographed an UNIDENTIFIED OBJECT on the Martian surface seconds before losing contact. Scientists described the UNIDENTIFIED OBJECT as a thin ellipse 20 KILOMETERS LONG! It was further stated that the photos could not be an illusion because it was captured by 2 different color cameras

as well as cameras taking infrared shots. One controller at the Kaliningrad control center concluded that the probe was now spinning out of control. It would seem that something struck or shot the *Phobos II* Probe.

That, it seems, was the official launch of *Phobos II* into the annals of UFO lore. Still, at that point, it was just a rumor, as no one had yet seen the alleged pictures of the 20-kilometer-long alien spacecraft. However, the photo was eventually made public in the West, and it is most likely just a shadow on the Martian surface cast by the moon Phobos itself.

Marina Popovich.

A few years later, in December 1991, A Russian test pilot named Marina Popovich joined the UFO chorus with an astonishing claim. Popovich, who was famous throughout the former Soviet Union as "The Russian Chuck Yeager" because she held 17 aviation world records, declared that *Phobos II* hadn't malfunctioned. She said it had been "taken out" by a UFO. As evidence, she presented a photograph she claimed was given to her by cosmonaut Alexei Leonov, who was the first man to walk in space and also a high official in the Soviet space program. Popovich said that she had smuggled the photo out of the former USSR against the new Russian government's wishes.

According to Popovich, the photo clearly shows a UFO hovering near the moon Phobos, and is "the first ever leaked account of an alien mothership in the solar system." She went on to declare that this object was apparently the cause of the disappearance of the *Phobos II* probe, and that officials in the Russian government were "alarmed" when they saw the image.

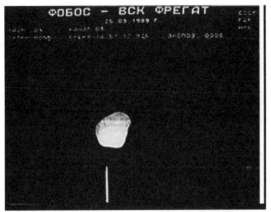

Last *Phobos II* photograph, released by Marina Popovich. Object in foreground is most likely a data artifact.

In reality, I think this is highly unlikely.

For one thing, Popovich is something of a UFO buff herself. In 2003, she published a book in Germany titled *UFO Glasnost* in which she claims that the Soviet military and civilian pilots have "confirmed" 3,000 UFO sightings. She further claimed in a series of lectures that the Soviet Air Force and the KGB have crash debris from five crashed UFOs, one of which was, according to her, obtained from the location of the 1908 "Tunguska event," which she attributed to the explosion of a nuclear powered UFO.

I cannot comment on these claims, other than to say they are presented without evidence. The Tunguska event has all the earmarks of a bolide air burst, which, while extraordinary, is a far cry from a nuclear detonation of an alien spacecraft. Such meteor explosions are rare, but not unprecedented, and are obviously of natural genesis.

Popovich's status as a UFO believer also makes her less

credible, in my opinion, because there is an inherent bias that comes with that territory. If she could provide sources for her claims, for instance a letter from Leonov confirming her story, it would be a big help toward confirming her information about what the Soviets privately concluded about the disappearance of *Phobos II*. Further, the Soviet Union was a communist state, and as such it was built on a web of lies and propaganda from the time of the Revolution. Such oppressive states have a vested interest in not telling their people (or the world) the truth. Masking the incompetence of their space

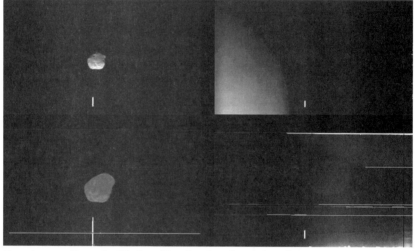

Four images from the *Phobos II* probe showing a data dropout camera anomaly (the author).

engineers by weaving a tall tale of a UFO encounter is exactly the kind of lie that such leftist regimes delight in telling. But there are even more reasons to conclude that *Phobos II* was lost due to a malfunction and not an interception by an alien "mothership:" The pictures themselves.

Along with the thermal infrared images of Mars, I also obtained a full set of the visual images that *Phobos II* took before its disappearance. As it approached the moon Phobos, *Phobos II* took a series of images of the moon itself. Some of them, but not all, contain what are known to be data transmission artifacts or camera anomalies. On four of the images, these data dropouts result in long lines of white and/or gray pixels. One of these, a vertical white line, appears in line with Phobos on several of the images, and changes length in all four of them. Under enhancement, there is absolutely no detail in these pixels at all—they are simply pure white. What this most likely means is that they are simply image artifacts, induced into the image either by the imager itself or in the process of being transmitted back to Earth. There does not appear to be a solid object there at all, but simply a data artifact. Based on this, I am forced to conclude that *Phobos II* was not shot down by a UFO, but simply malfunctioned due to an error by the Soviet engineers. Most likely, the UFO story was leaked (along with cleaned up pictures) to certain people in the Soviet space program that were known to be talkers, and they took the bait and leaked the UFO story, neatly covering up the Soviet engineers' costly and embarrassing mistake.

So much for the UFO theory…

But as we waited for new THEMIS images of Cydonia, the *Phobos II* data was never far from my thoughts. While the UFO shoot down scenario had been ruled out, we were still left with the extraordinary thermal IR data and the images of Phobos itself. Things began to heat up on the THEMIS end of things when word began to leak through Space.com reporter Leonard David that early science data from THEMIS was described as "amazing" and a "whole new Mars" by NASA scientists.[2] NASA then scheduled a press briefing for March 1, 2002, 13 years to the day after *Phobos II* had taken its extraordinary thermal infrared image of the Martian city in 1989.

Based on NASA's ritualistic pattern of using certain dates over and over again to release data and land spacecraft (see *Dark Mission*) Hoagland and I predicted on his *enterprisemission.com* website that the images shown at the press briefing would be thermal infrared and that they would be of the Hydraotes Chaos region.[3] This was the same area that *Phobos II* had imaged in the thermal infrared 13 years before.

As it turned out, we were dead right. The assembled scientists, which included the THEMIS program manager Dr. Philip Christensen, showed a number of images and presented data which showed the distribution of water on Mars. That data confirmed our Mars Tidal Model predictions. They also showed the first THEMIS thermal infrared image, and as we predicted, it was taken of the Hydraotes Chaos region. The image showed large, block-like "mesas" that were geometrically arranged (like a city) and which had highly unusual heat retention properties. They also had rectangular notches, ground level "openings" and many other unusual characteristics. All of this, however, went uncommented on by Christensen *et.al.*, and the assembled press didn't ask. But we were about to get more—far more, it turned out—from Dr. Phil Christensen than we had ever expected. It was all about to get—to quote Morpheus from *The Matrix*—"a little weird."

(Endnotes)

1 http://themis.asu.edu/feature/42

2 http://www.space.com/science-astronomysolarsystem/odyssey_update_020226.html

3 http://www.enterprisemission.com/expect.htm

Chapter 2
Mars Heats Up

As the spring of 2002 drifted into summer, the independent researchers were more optimistic than we had been in 20 years that we were on the brink of a breakthrough with Cydonia. *Mars Odyssey's* chief project scientist, Steven Saunders, had made a number of positive statements to members of Hoagland's bulletin board service regarding the possibility of getting multi-spectral color images of the area around the Face. Saunders then announced that the Cydonia infrared data that we had been seeking for more than a year would be released a few days later, in late July 2002. The independent research community knew that if this dataset was honest, we might possibly have a "smoking gun" that would end the debate about Cydonia's artificiality once and for all.

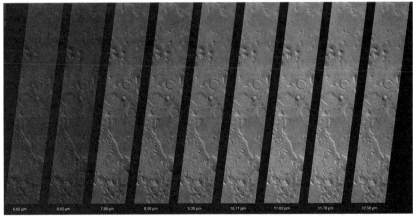

Themis thermal infrared images of Cydonia (ASU)

Because of the ritualistic way in which NASA does so many things (see *Dark Mission*), we expected and predicted that the image would be released on or about July 25th, the date of the release of the first "Face on Mars" image, 35A72, back in 1976. As we expected, the image was finally released on July 24/25, 2002, after midnight on the east coast but still on the 24th JPL time. This made it exactly 26 years to the day—if not the hour, from the release of 35A72. But it was a disappointment. Instead of the full five-band color, fully processed multi-spectral image we had expected, we instead got a series of nine grayscale image strips, lined up side by side on a black background canvas. Even so, Hoagland, and I and others downloaded the data and began to work with it. Because of the potential controversy these images might cause, Hoagland and I then decided to stay completely out of the initial "processing phase" of the THEMIS infrared images. We instead elected to rely on two outside volunteers in image processing who had freely offered technical assistance: Holger Isenberg and Keith Laney. Isenberg was a graduate engineer for Applied Computer Science at Dortmund University, as well as a Unix System and Network Administrator at a German software company. Keith Laney was a digital imaging and software applications specialist and Mars Orbiter Camera (MOC) image processor for the NASA-Ames MOC MER2003 Landing Sites Project. It was believed that using these two specialists would give the maximum degree of credibility to the eventual results. Since the other key factor in gaining credibility in scientific circles is reproducibility, we also decided that Keith and Holger should aim at accomplishing the same tasks, but work as separately from each other as was practical.

Keith, at first, was not very enthusiastic about any aspect of the project. Hoping to see a full-color, multi-spectral image (per Dr. Saunders' preview web announcement), Laney was so displeased with the grayscale, side-by-side, multiple-band image strips on the THEMIS website that he initially didn't even bother to download it. Flatly declaring that "it sucked" when asked about it on various message boards across the internet, Laney went back to working on Mars Orbiter Camera visible light images.

Because of this highly visible position, Keith (and a few others who had also publicly criticized the quality of the new image) then began to get a series of interesting responses on the bulletin boards and in private chats. These responses came from two new visitors to the boards: somebody calling himself "Bamf," and another person named "Dan Smythe." Bamf had begun posting about a month before the image release and Smythe appeared simultaneously with its release. Both of these newfound "friends" seemed to know quite a bit about infrared image processing, and harassed and goaded Keith Laney into finally going back and downloading the full-size, uncompressed TIFF file—which according to his own computer log he did on July 25, at 10:27 p.m. Eastern Daylight Time. This timestamp will be important later, so make a note of it.

We were all from the get-go more than a little suspicious of both Bamf and Smythe. That suspicion only deepened when we discovered that "Bamf" was in fact none other than Noel Gorelick—who claimed in his on-line profile to be the manager of Arizona State University's Mars Computation Center. In other words, Gorelick was ASU's Dr. Philip Christensen's right-hand man. Why exactly the right-hand man of the chief investigator for the THEMIS instrument was doing lurking around Hoagland's message board was not immediately clear, but Gorelick went on to further claim that he was in fact the person who had hand-assembled (over a period of four days) the infrared Cydonia image posted on the THEMIS website.

From the beginning, Gorelick was publicly acerbic and disparaging of the very reason for the existence of the message board, which was the discussion of the possible artificiality of the Cydonia anomalies, like the Face. But privately "Bamf" and Smythe provided Keith and Isenberg—and anyone else with the smarts to work with the new image—a virtual on-line "how-to" tutorial in infrared image processing. Given that Gorelick was essentially nothing more than a webmaster, it was unclear where he got the expertise to guide Keith and Holger in their infrared image processing. More likely, it seemed at the time, the technical expertise was being provided behind the scenes by Dr. Christensen himself using Gorelick as an intermediary.

Using the advice he'd been given and the image he had downloaded on the 25th, and working with a state-of-the-art software suite that included ENVI 3.5, which provided a crucial "decorrelation-stretch" tool, Keith began to achieve astonishing results once he got the hang of it. His initial color ratio composites—

from the nine strips of drab B and W imaging he had downloaded from the official THEMIS website—were nothing less than stunning, even in 8-bit grayscale.

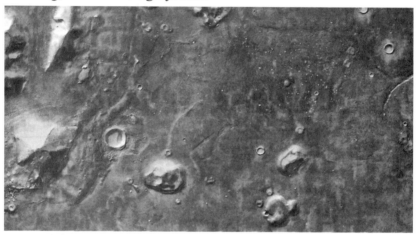

Processed thermal infrared image of Cydonia. Face is to the right and the D&M pyramid is center-middle. The "Fort" is partially visible at the top center.

In addition to a whole new perspective (in the infrared) on the long-familiar features of Cydonia—the Face, the D&M, etc.—Laney found a troublesome, quasi-regular pattern of easily discernable block-like markings running almost the entire length of each band of the IR Cydonia strips. When he queried Gorelick about the annoying "blockies," as he called them, "Bamf" grudgingly admitted in a chat, "so, you've found our dirty little secret." The ASU web manager acknowledged that the THEMIS team had also seen this peculiar pattern, and had assumed it was digital "noise." It was because of this pattern, he claimed, that the ASU team had decided not to create the normal, multi-spectral color images produced of other areas on Mars from the same THEMIS camera. Remarkably, he also admitted that, despite their best efforts, they could not seem to remove the "noise"—or find it on other published IR THEMIS images of any other region of Mars. It seemed it was unique to the Cydonia THEMIS IR images.

Keith then began feverishly working on his own methods to remove this offending pattern, even sharing a couple of de-striping tricks with Gorelick (who was quietly soliciting outside assistance on the problem) in the process. But, ultimately, he had no more luck than the official THEMIS team.

Around the same time, Hoagland, who was following this exchange between Bamf and Laney via e-mail, decided to suggest a radically different process for eliminating this peculiar "noise." Having worked in science for decades, he was aware of a technique for noise reduction (or averaging) used in astronomical photography called "luminance layering."[1] It involves adding a visible light image to an existing thermal infrared or near-infrared filtered image. This results in a reduction in graining, an increase in contrast and detail and a much superior overall image. Laney had decided independently to try the same thing on his own even though he had never heard of luminance layering. Once they were agreed

Luminance layered THEMIS thermal infrared image of Cydonia. Face is in the center and "Fort" is to the lower left. The block pattern is not a compression or processing artifact but is an actual feature of the Cydonia region just below the surface.

and both Keith and Hoagland added the visible light image to the color stretched IR data, the results were breathtaking.

What both men found to their surprise was that the pattern not only wouldn't go away, it actually got *more pronounced*. It suddenly struck Hoagland that what he and Laney (not to mention the THEMIS team itself) thought was "noise" in the Cydonia images might actually be the *signal*. It was then they agreed to switch gears and start trying to *amplify* the pattern in the ongoing ENVI processing of the raw data, rather than continue efforts to remove it.

At the same time, taking the composited VIS/IR "decorrelated" versions Laney had already completed, Hoagland began a few experiments of his own. That's when the "cityscape"—miles upon miles of individual, clearly artificial structures hiding beneath the dusty plains of Cydonia—became completely obvious.

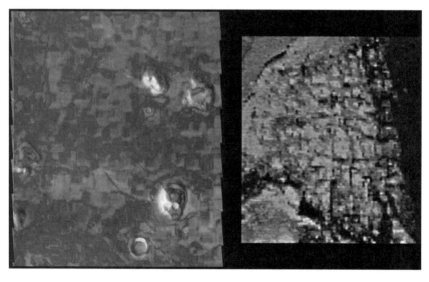

By simply increasing the image saturation of the composite VIS/IR image, what had been an annoying and mystifying block pattern was suddenly revealed to be a stunning, highly detailed *city*, silently sleeping for untold ages beneath the timeless sands of Mars. Looking very similar to the block pattern imaged by *Phobos II* decades earlier, it quickly became apparent the cityscape was real and not some image artifact. For one thing, the pattern did not follow the scan lines of the imager, but rather true Cydonia north and south—exactly as terrestrial cities most often do. In addition, there were a wide variety of individual "structures" within the pattern which did not conform to this north/south grid at all. This made us completely confident that these structures are not noise, since they show, in both layout and design, clear evidence of architectural intent and organization. And even the smallest structures were at least an order of magnitude (ten times) above the very low noise threshold of these superb VIS/IR composites, in terms of resolution as well as intensity.

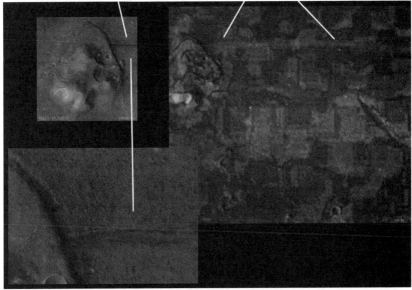

The Fort.

Without any further enhancement, massive individual structures literally fell out of the data by simply turning up the color saturation. Previously identified features, like a "tunnel" running out from under the Fort, could now be seen to continue out across the Cydonia plain and under the soil. The ground penetrating capability of the thermal infrared images showed that the exposed portion seen in the visible light images was only a small segment of a much longer tube structure that ran just underneath the visible surface from the Fort for several miles before it terminates in another structure. However, in the visible light images, the tunnel runs for only a few hundred feet before it disappears beneath the Cydonian sands. The truth was, it looked for all the world like a subway tube you might see in London or New York or any number of other cities on Earth, connecting one part of town to another.

Close-up showing miles long subsurface tube-like structure extending from underneath the "Fort." Note the bright "Temple of Isis" structure in lower middle and city block like layout.

The close-up views of the tunnel revealed that it had a distinct rib cross-structure—strikingly similar to the other Martian "tunnels" found by Hoagland two years before in a *Mars Global Surveyor* image. The image showed a series of regular, repeating ribs which encased the tube and seemed to be a support structure. The image was so striking that science fiction writer Arthur C. Clarke placed the image on his website and declared that NASA has found fossil evidence of life on Mars.

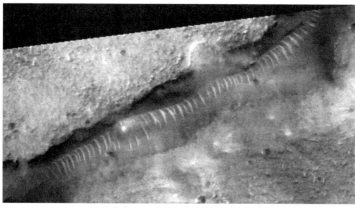

MOC image MO-400291(NASA).

Our take (Hoagland's and mine) was that more likely than a giant sandworm skeleton, like something out of Frank Herbert's novel *Dune*, this and other similar objects which began popping up on various *Global Surveyor* images were some sort of exposed, subsurface transport system, possibly made of glass. NASA's explanation, that these were simple "dune trains" came without any images of comparable examples on Earth, and was quickly dismissed by independent geologists. The transport tube idea was supported by the fact that the ribs actually *wrapped around* the underlying tube structure, which simple sand dunes just don't do. In addition, the whole structure was translucent and yet highly reflective. There even seemed to be some kind of object *inside* the tube, visible "behind" the ribs. Add the brilliant and completely inexplicable specular reflection coming from the encased object (a train car?) and we felt we had a very strong case. To now find another such tube in the thermal infrared data extending under the Martian surface only reinforced our certainty. Later, the late astronomer Dr. Tom Van Flandern joined the chorus and commissioned his own view of what this transport system might have looked like in its heyday.

But was this simply the return of the Martian canal builders? A closer look at the subsurface city of Cydonia revealed in the thermal infrared showed that there were multiple examples of these tube-like structures, buried perhaps just a few meters below the Martian sands.

Artist's concept of Martian tube systems.

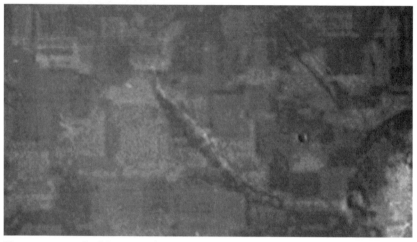

Transport tunnel with supporting cross members running from the "train station" in downtown Cydonia to the base of the symmetrical mesa.

One new transport tunnel emerges from the base of a building we are now calling the "train station" in downtown Cydonia and runs right underneath a symmetrical mesa just south of the Face. In fact, it intersects the mesa at the exact lateral center of that massive above-ground structure. This is exactly what one would expect from an artificial transport system connecting to artificial structures. Close examination of the other end of the tube shows that the tunnel terminates at the base of the "train station" directly in front of a series of striking open archways in that building.

What was absolutely clear from the new THEMIS enhancements is that these various "tunnels" are indeed just that. The THEMIS image allows us to see that the tunnels running out

Close-up of the "temple" (left) and the "train station" (right). Note bright tube-like structure emerging from archways in the train station and cross-supports at 90 degrees to the tube.

from the Fort and the train station are not just simple dune filled depressions in the landscape, but coherent, tubular structures that run for many miles beneath the Martian desert and across the Cydonia plain into the heart of the buried Cydonia city. There are also clear structural cross-members—massive in scale—that run 90° to the tunnel at various points along its traverse, as well as its companion tunnel leading to the mesa south of the Face. Architects and engineers will again quickly recognize these features as obvious evidence of intent and design. The simple fact that the tunnels can be seen to begin above the surface and then continue below the ground is clear evidence that they are not natural formations of any kind.

And there were the other, even more unimaginable wonders on this THEMIS image, the most prominent being a stunning, overwhelmingly architectural structure we called the "Temple."

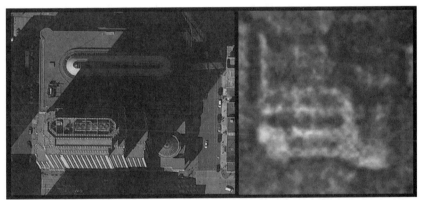

The "Temple of Cydonia," right, and a terrestrial comparison, Minneapolis, Minnesota (Bara\ASU).

You can clearly see, without even straining your eyes, the individual cells in the roof and what appear to be decorative buttresses all around this multi-story structure. This object, just east (to the "Face side") of the Fort, is an architectural marvel—clearly a structure of significance and, with the exception of its apparently missing roof, in nearly pristine condition. It is also enormous, taking up an area the size of an entire city block all by itself. We quickly detected numerous other completely obvious buildings of various types scattered all across the image.

Obviously, we were excited by our new find. This data fit exactly the kinds of descriptions we had been getting from inside sources, and even NASA's Dr. Saunders. It was very tempting to release the data immediately—but before we could go public, there were a few issues to resolve. First, if there truly was an underground city at Cydonia, how did it get there? Was it older or newer than the eroded ruins above the ground? And even more pressing, how could we see it at all? IR images had some degree of ground penetration capability, but these images seemed to be visualizing objects well beyond the published capabilities of the camera.

Longwave infrared imagers have a limited ability to penetrate underground by detecting subsurface "hot spots" conducting heat upward from below. But, they typically cannot show the kind of detail we were seeing. One possibility was that we were remotely sensing our "city" through a layer of fluffy, micron-sized dust essentially transparent to the THEMIS wavebands; probably covering another, thicker layer of protective and also infrared transparent material, most likely ice.

Exactly how that ice got there, and how we were seeing the city below it, quickly became secondary considerations, however. Just as we were ready to go to press, we suddenly became aware that the original THEMIS daytime IR image of Cydonia posted on the website may have been altered.

The first hints came when Holger Isenberg, our imaging associate working from Germany, was unable to duplicate our key results. Holger was not working with precisely the same software suite as the rest of us. This was both a good and bad development, as it inadvertently gave us a chance to compare the results when two different but functionally similar sets of imaging tools were applied to the same IR data.

The first assumption was that there might be something that either he or we were doing wrong that gave us differing results. We eventually discovered that Holger was using a PNG version of the THEMIS image, as opposed to the full size TIFF (Tagged Image File Format, a computer image format without compression) that the rest of us were working with—but that did not account for the entire discrepancy that we were seeing. PNGs are a fundamentally solid file format, completely lacking the "lossy" compression problems of JPGs and GIFs.

Beyond that, the objects that we were finding were so big, so far beyond the noise level of the image and so easy to bring out that even with a compressed web image file format, Holger should have seen these objects literally jump out at him. Instead, there seemed to be a distinct lack of information/image data in the images he was processing.

And all of this came on the heels of the dramatic and curious developments of the preceding week or so. Our newfound friend, Bamf, aka Noel Gorelick of ASU, made some emphatic statements that seemed to fly in the Face of his previous attempts to help our efforts. On or around August 20, 2002, Gorelick made what seemed at the time to be a bizarre and arrogant claim. He insisted he had the power—and indeed the right—to flagrantly alter the data posted on the official THEMIS website… at his whim. To quote him verbatim:

> The daily images website is a service I provide and support entirely because I'm a nice guy. There is no NASA mandate or contractual requirement for me to provide these images to anyone before they're delivered to the PDS [Planetary Data System]. I do so because the public is interested in what's going on with the mission, and it's good for public relations. Accordingly, if I feel like degrading the data before I post it, I'm certainly free to do so. If I want to scribble on the images with a crayon before posting them to *my* website, I'm probably free to do that too. The 'government data' that the public paid for is being well cared for while it's being prepared for delivery to the PDS.

Tagged on to an earlier, equally strange statement in the same vein made by Gorelick—claiming that "no one can tell if an image has been fundamentally altered"—we began to become more than a bit concerned. This, in turn, led to us withdrawing from our original plan to post the infrared Cydonia THEMIS data and our preliminary analysis on Hoagland's website on August 22, 2002 while we rechecked our data's fundamental integrity and lineage. Hoagland announced this on Art Bell's *Coast to Coast AM* show that night.

That prompted an immediate response from ASU's Dr. Philip Christensen, the principal investigator for the *Mars Odyssey 2001* THEMIS instrument. He sent an e-mail to me late in the evening on August 22, 2002, as if he had been waiting either at home or in the office for Hoagland's appearance on *Coast to Coast AM*:

> Dear Michael,
>
> I am confused by your statements regarding the THEMIS IR data and your decision not to release your findings. The data were calibrated [sic] our standard processes and in the same way that is done for the THEMIS science team. I am not sure why you are suggesting that Noel, or anyone else on the THEMIS team, has done anything to alter the data—he was simply questioning how you have treated the data and how you have validated your methods and processes.
>
> I look forward to your release and to a detailed description of exactly what you have done to the data once you have downloaded [sic] from our website. Hopefully you will provide an accurate description of these techniques and the methods that you have employed to validate everything you have done to the data.
>
> Sincerely,
>
> Phil Christensen, THEMIS Principal Investigator

At this point, Hoagland, Laney and I made a conscious decision to pretend in public that we were more than satisfied with this response. But the truth was that it only deepened our concern. Not only does Christensen's e-mail make a blatant misstatement— Gorelick certainly *did* imply he might have altered the data (and more than once), and was not simply "questioning how you have treated the data." If threatening to "scribble on the images with a crayon" doesn't strike Christensen as a reason for us to infer that Noel Gorelick might have done something untoward to the data, well, Phil, what would qualify?

The most disquieting part of the e-mail was Christensen's inordinate interest in our "methods and processes." Why was he so interested in what *we* had done to the data, when he himself and his own team had provided nothing to the public, Gorelick's assertions to the contrary, about what *they* might have done to it? The THEMIS team hadn't provided any description as to how the data had been handled, treated or possibly enhanced. They hadn't provided the slightest hint of what filters may have been applied, a paper trail of who had handled it, or even the most rudimentary ancillary data (spacecraft geometry, lighting, etc.) for any of the released *Odyssey* images. If they weren't even willing to provide the date, time and orbit number of the THEMIS Cydonia image, why would they be so interested in *us* providing an "accurate description" of what *we* had done?

We didn't quite know yet, but we knew we needed to find out—so we went back through the data stream and tried to find any discrepancies in either our processes or tools. The one thing that kept coming out was that Holger Isenberg just did not seem to be getting

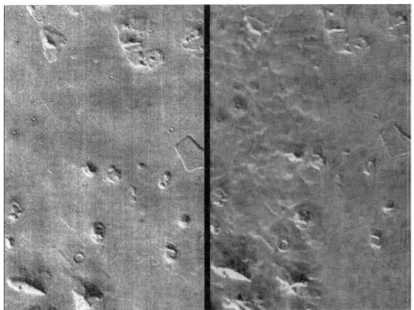

Processed "July 25th" version of THEMIS IR Cydonia image (right) vs. identically processed "August 25th" version downloaded from the *same* ASU web link (left). Later image has block-like structures erased and excessive noise deliberately introduced.

49

the same results as Keith Laney. In fact, he wasn't even coming close. Granted, he didn't have full use of the ENVI software, but that alone didn't seem to fully address the striking differences. At this point our "separate but equal processing strategy" seemed to have effectively run its course.

Keith was nonplussed by this problem as well. After initially thinking that Holger just wasn't getting it, he finally decided to retrace his own steps, go back to the THEMIS website and download the image one more time and start from scratch, working with Holger step by step. When he did so, he—to put it bluntly—got a major shock.

When he brought up the two images side by side—the one he'd downloaded on the evening of July 25, and the one he had just re-downloaded from the exact same official THEMIS website on the evening of August 25—they didn't match.

As you can clearly see from the comparison graphic, Keith's original July 25 version is noticeably different—even to naked eye inspection—from the one currently posted (as of this writing) on the official THEMIS website. In Keith's version, you can plainly see the block pattern even in the so-called raw data, without any computer processing. In the official THEMIS posting, these blocks are completely absent. The differences become even more striking when standard IR enhancement processes are applied. In Keith's original version, he could easily replicate his previous results (with ENVI) and get down into the data. The result is a beautifully clean IR composite image with very little noise. When he used precisely the same processes on the official August 25th website version, the result was absolute garbage. The newer replacement image had the blocks (the individual subsurface buildings) erased and an excessive amount of digital "noise" deliberately introduced.

As the ENVI software manual points out, in a true multi-spectral IR image, you cannot destroy or add to the data with a simple filter, like a Gaussian blur. You can only enhance it, because each individual IR pixel retains the same base intensity data that it had originally. On the other hand, if you reduce the resolution with resampling, overlay a visible light image and then apply a sharpen filter (while adding an unknown amount of pure white noise)—as apparently had been done with the official website version—then you will get pretty much what we see on the degraded "official" version.

Or, to paraphrase my old buddy Stuart Robbins, you can tell that one image is manufactured from another by the amount of noise in the respective images. Given that the later, "official" version has far more digital noise in it, the obvious conclusion is that the later image is a degraded version of Keith's original. In other words, a highly diluted fake.

Some researchers argued (nonsensically) that because Keith's image was clearer and showed unmistakable individual buildings and structures on it, that it must be the fake. They insisted that the THEMIS IR camera wasn't good enough to resolve objects that clearly which were below the visible surface. This is simply not correct. Indeed, if you read the abstract from one of Dr. Christensen's own papers on the THEMIS results, he makes it quite clear that the instrument is more than capable of achieving excellent "ground penetration." To quote from the abstract:

> Regional 100m mapping has revealed the presence of channel systems in ancient crater terrains not detected by Viking and not mapped by the high-resolution camera on Mars Global Surveyor.

In other words, "We're seeing stuff with the THEMIS instrument below the ground that can't be seen on the visible light *Viking* or MGS images."

Faced with this discrepancy, we elected to hold off publishing on Thursday night, August 29, 2002, and instead simply put the source data out side by side so that it could be evaluated by the public. We waited to see if there was a response from Christensen, but when none was forthcoming, we elected to take the initiative and begin an e-mail exchange, which was also ultimately fruitless. In the absence of absolute proof one way or the other, we were left to go over the points which make it overwhelmingly clear to us that the image Keith was working with was the real deal. Our analysis resulted in four separate *proofs* that led us to this conclusion. In addition, there are several other, finer points that reinforce our stance.

First, we know that Keith's image and the official image probably don't come from the same source. The header tag in Keith's version shows that it was converted from a PNM format, a standard conversion format used inside NASA to process remote probe images. In Holger Isenberg's opinion (remember, he's a UNIX administrator for a German software company), Keith's image was also processed on an older UNIX-based computer—typical of the type of equipment used by universities and the government at that time. The "official" version, on the other hand, appears to have been processed on a Windows-based machine and shows no sign of having been converted from a PNM. So, there is nothing inconsistent in the header tags with Keith's image being the genuine article.

Since we are all used to working with data in the visual light spectrum, we have been taught to assume that sharper is better, which it is—if you are working in the visible light bands. However, visual resolution does not equate to infrared sensitivity or information richness. The real image, while it may look a bit blurry, in fact contains far more data than the official version, no matter what the official version looks like on first (naked eye) inspection. That's why the processing and enhancement tools used for multi-spectral

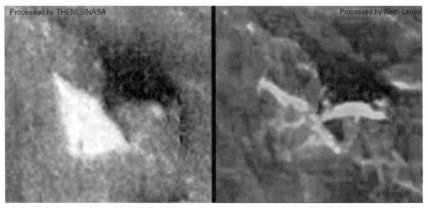

The D&M in infrared. Note far greater amount of noise in the "official" version, left.

imaging, to accomplish exactly that, are crucial to extracting that signal from this set of IR images.

When you do a visual inspection of the two sets of raw data side by side, there are additional visual clues that will tell you that the "real" version is true THEMIS IR data, and that the "official" version is at best a poorly rendered copy. The first proof of this can be found with a simple visual inspection of some specific features in the raw versions of both the "real" and "official" versions of the THEMIS Cydonia IR data.

In Keith's real version, there are subtle but distinct differences in specific features from frame to frame. This is what true IR data should show, because each signal return in each individual IR waveband is going to find a slightly different "bottom" on the planet below. The longer wavelength signals will penetrate more deeply than the shorter wavelength signals. As an inevitable result, there will be subtle changes in the appearance of certain features—but only if they are "real" features on or just below the planet's visible dusty surface.

Conversely, in the "official" version of the Cydonia IR dataset we found no difference whatsoever from frame to frame as you go up the infrared bands. All you can see is steady, overall brightening. This proves that the top layer of the official version is not only not real infrared data (it's a visible light image), but that someone must have simply brightened the entire image, or methodically darkened the individual image strips, in an attempt at misdirection—a blatant fraud made up to look like real IR data to the uneducated eye.

The second major proof validating the real version is obtained by a comparison of specific features in earlier visible light images of Cydonia. There are numerous areas on the *Odyssey*, MGS and *Viking* visible light images that can be inspected and compared to the real IR data. When we do so, we would expect to see some of the block features that appear in the infrared to also appear in the visible light images, assuming that some of these buried structures are actually near the top of the dusty, icy layer.

There are numerous examples on MGS visual images of block-type features matching up with the IR blocks. Hoagland posted several on his website at the time. If these were processing artifacts (or scanner marks, as one person actually suggested), then there would not be *any* correlation between visible features and the IR blocks. The existence of even one correlation constitutes a *proof* that the blocks represent real features on or just beneath the Cydonia dusty plain.

The next proof is in the analysis of the noise floor.

Using an ENVI-generated data table comparison, Keith's "real" version was much clearer, showed far more detail, and simply contains far less noise than the official version. The "real" image had a much wider range of spectral data in it, owing to the enhancement capabilities of the ENVI 3.5 decorrelation-stretch technique and the much cleaner source data. The official version may have actually been generated from the same original, but the spectral range of the image—in essence "the signal"—was deliberately compressed, resulting in the extremely noisy and scientifically worthless "official" image. This second major point—which numerically validated for us the entire reality of Keith's version—is that it is entirely consistent with the kind of quantitative results one would expect from the exquisite THEMIS instrument, while the "official" version is entirely *inconsistent* with what that amazing instrument is capable of producing.

According to a JPL document (which was subsequently removed from the JPL website after I downloaded it), the THEMIS instrument is accurate to plus or minus 0.001K (one thousandth of a degree Kelvin) temperature measurement. That means that it is capable of differentiating temperature differences—instrument thermal "noise"—in increments that are incredibly precise.

What that translates to visually is brought out when you decorrelation-stretch the images—as Keith did with ENVI 3.5— essentially separating the thermal data from the compositional data to allow you to see more detail in the "composition bands." And that is why, when we added the visible light overlay to the "good" THEMIS data the screen literally exploded with rich detail and stunning clarity. In short, Keith's version of the THEMIS data behaved exactly as real THEMIS data should—and the "official version" posted on the ASU website did not.

The final and most conclusive proof that Keith's version of the Cydonia IR image is the real one comes, from of all places, the very instrument on another spacecraft (MGS) that was fraudulently

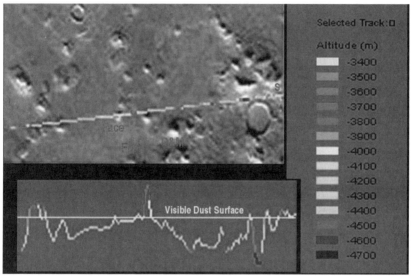

Mars Orbiter Laser Altimeter data showing actual terrain of Cydonia.

used by NASA in a failed attempt to debunk the Face back in 2001. Published MOLA (Mars Orbiter Laser Altimeter) data from *Mars Global Surveyor* can be used to ultimately clinch the "buried city" model.

As I discussed in *Ancient Aliens on Mars*, there was at that time only a single MOLA pass over the Face on Mars and the Cydonia complex. The MOLA instrument operates at the 10.6 micron wavelength, essentially smack-dab in the middle of the THEMIS sensing band. The altimeter is not capable of penetrating solid ground at that wavelength, but it should not have a problem piercing a fluffy layer of micron-sized dust and an underlying long-wave, infrared translucent material, like ice. When we graphed the altitude scan, we not only found that that the majority of the ground truth at Cydonia seemed to lie *below* the mean datum of the visible plain, but in some cases it was many *thousands of feet* below. In looking at the visible light images, Cydonia was a flat and fairly smooth plain. But what the MOLA data told us was that beneath a thin layer of dust, it was actually an area full of very deep toughs, certainly deep enough to house the entire "City" that we were seeing in Keith's infrared images. It flatly confirmed that Cydonia is not a flat, relatively featureless plain with some mildly eroded mesas scattered about. It is, in fact, a deep, dust-covered plain which

Cydonia City.

lies atop a thick layer of ice that has preserved and concealed—literally for countless millennia—a once highly advanced, plainly technological and now mysteriously absent civilization. All it takes to prove this is to go there, brush away the dust and start drilling.

There can be no middle ground here. Architects and engineers will instantly recognize the form, fit and function of their craft in the objects scattered all throughout the THEMIS image. And even more, some of the features can be seen, just barely covered over, in the *Odyssey* visible light image and even on the original *Viking* data. How "noise" could occupy the same area and location on images taken 25 years apart, by completely different instruments and on completely different spacecraft, is going to be a tough hurdle for the usual suspects to clear. In addition, the MOLA data confirms that a great deal of Cydonia exists *below* the visible surface precisely in the basin occupied by our frozen city. The laser can't lie.

So Keith's version of the IR image—the only one which is

consistent with all the other observations—must be the closest thing to the "real" data returned by the THEMIS instrument we have now in the public view.

There is one final point. We were now forced to reluctantly conclude, based on this "bait and switch" maneuver apparently cooked up by someone inside ASU's official THEMIS team, that someone in NASA/ASU was trying very hard to set us up. Had we gone forward with our original announcement and revealed the images without the background story concerning the two conflicting datasets, we would have naturally called upon NASA professionals and experienced IR imaging experts in the private sector and academia to replicate our work. They would have then gone to the official THEMIS website, downloaded the degraded "official" version of the THEMIS IR data—and been totally unable to accomplish that, even given the proper multi-spectral software and experience. We would have been thoroughly and publicly discredited, certainly in the eyes of the curious middle of this debate, and any nascent interest in the possibility of a former, now-extinct Martian civilization beginning to stir among some of the honest professionals inside NASA or in the press would have quietly died, as someone clearly intended.

What we did not know was why, nor did we have any concept of the psychological reaction our exposé of NASA's duplicitous behavior would have on the independent researchers that were supposed to be on *our* side…

(Endnotes)

1 http://www.robgendlerastropics.com/LRGB.html

Chapter 3
BAMF! You're it!

"The first indication of intelligent life on Earth lies in the geometric regularity of its constructions..." —Carl Sagan

Comparison of THEMIS Cydonia thermal infrared image (top) and farmlands on Earth (bottom).

At this point, we knew a few things for certain. First, the incredible geometric regularity of the processed thermal infrared data was an almost indisputable smoking gun for artificiality at Cydonia. While it raised new questions—for instance, was the underground city older or newer than the ruins above?—It also confirmed long held

59

suspicions. Not only was Cydonia a region of massive artificial construction, but it was now clear NASA knew this and were willing to go to great lengths to make sure the general public did not accept this as an axiomatic conclusion. We also knew that someone inside JPL had tried to set us up, tried to sucker us into releasing data that couldn't be confirmed independently by using the "official" data. We knew that Keith's data was real, because it had passed every test and was entirely consistent with what THEMIS data should look like. What we didn't know was just who had been behind this rather clumsy and desperate attempt to discredit us.

In order to separate the good guys from the bad guys, our first order of business was to determine just how and why Keith Laney was apparently specifically targeted to receive the original data, and just who pulled the necessary strings to implement this part of the plan. In order to reconstruct that trail, we had to go back to the two main players in this little psychodrama, "Bamf" and "Dan Smythe," who we now strongly suspected was Dr. Philip Christensen's on-line doppelganger. Both had shown up on Hoagland's discussion forum within about a month of the release of the Cydonia IR data. This was shortly after Hoagland's apparent breakthrough in communications with NASA's Dr. Jim Garvin, and before any of the independent researchers had a hint that the Cydonia IR data release from *Odyssey* was coming.

In retrospect, it is now obvious that was a reconnaissance mission, an attempt to sort out just who among our crowd might have the knowledge and expertise to work with the real data once it was released. Keith Laney would be an ideal candidate for this kind of operation. He was reasonably independent of Hoagland and me, worked for the NASA/AMES *Marsoweb* project, and had quite a bit of experience in visual image processing. What he lacked was any experience working with multi-spectral data. Into this void stepped Gorelick and Smythe, ready, willing and able to train him.

This all set up perfectly for the release of THEMIS thermal IR data on July 24, 2002. Keith Laney, almost certainly their number one candidate to receive the "real" IR data, took a quick look at the public release, blurted out "it sucks" and didn't even bother to download it. Hoagland quoted Keith that night on *Coast to Coast AM* and, following that, Keith got into a series of pointed message board exchanges with both Bamf and Smythe (Gorelick and Chrsitensen). Eventually, after much prodding by them, Keith went back to the THEMIS website and downloaded the image we now know is the real image, on July 25, 2002 at 10:27 p.m. EDT. In fact, as we studied the logs of the discussions, Smythe and Bamf aggressively badgered Keith into going back to the site and downloading the image.

As Hoagland and I wrote about in *Dark Mission*, NASA has a habit of following certain ritual dates and numbers, most prominently 19.5 and 33. It did not escape our notice that the release date of the new *Odyssey* IR data was exactly 31 years *to the day* from the first image ever taken of the Face on Mars by *Viking I* on July 25, 1976. In the course of processing the data, Keith Laney also discovered that when you crop and rotate the individual image strips (remember, created by Bamf/Gorelick) to the vertical, and examine the pixel count, it turns out that the image strips are exactly 1947 pixels by 333 pixels, down and across. Or, 19.5 x 3.3, if you are tracking this sort of thing. We took this as yet another hint from Gorelick that this was the true Cydonia data that they had been working with on the inside and that had so amazed the JPL scientists in the Space.com articles.

Once Keith had downloaded the real image, Gorelick proceeded to provide enough assistance to Keith (and Holger, among others) to enable Laney to eventually figure out how to properly process the data he'd been given, and exactly what software he would need to carry out the job. Gorelic interlaced this crucial private e-mail tutoring in multi-spectral imaging with his previously mentioned bizarre public comments on our investigation, along with other categorically false assertions on the THEMIS Cydonia data. Among these were unambiguous statements that "the IR data on the website were useless" because they had "not been calibrated," and that "the Face is not any different than any other object at Cydonia... is not made of anything unusual, like metal or plastic..." Not only are both these statements wrong, they are also flat out contradictory.

A JPL paper on decorrelation-stretch techniques flatly states that calibration is not necessary to make use of decorrelation-stretched IR data, because of the nature of the algorithm and the quality of the process itself. All calibration ultimately does is identify which colors relate to which materials—assuming you know what materials you are dealing with in a totally alien context in the first place. There is, obviously, some question on that matter raised by the Cydonia data.

And as for Gorelick declaring that he "knew" what the Face is composed of after he already claimed that the data "wasn't calibrated," this simply made no sense. Forgetting that it conflicts with not only the image caption that he posted, it also conflicts with Christensen's e-mail in which he restated that the data *was* calibrated. So for Gorelick to presume he knows what the Face is made of, when by his own claim he is working with uncalibrated data, was and is nonsense. And, even if the data *is* calibrated, as the THEMIS website and his boss now claimed, he still couldn't possibly know what the Face or anything else is made of, because as of the moment, nobody knows what "Cydonia City" is composed of. Unless of course there is a special library on alien construction methods and materials available inside JPL.

All of these conflicting signals left us with a few certainties, but many more unanswered questions. We knew Keith's dataset was valid, and we knew that Gorelick had tried to help us to some degree. We also knew that somebody was hoping to sucker punch us by changing the data on the official site, in order to create profound confusion when we tried to have our own analysis verified by other independent researchers.

So this left us with two questions: just who were the good guys and bad guys inside the ASU/THEMIS team, and how did they get the right data to Keith Laney—and only Keith—for their ritually perfect July 25th window?

What then became clear was that the answer to one question should provide the answer to the other. Keith Laney had a cable modem connection to the internet. What that means is that he has a "static" IP address. Every time he goes on-line (or even has his computer turned on and physically connected to the web), he broadcasts the same identifier to the whole world. It would certainly be a piece of cake for any reasonably savvy computer network expert to "ping" Keith's system, determine his static IP address and lay a trap for him.

Once Gorelick and Smythe had convinced Keith to go back (on the 25th) and finally download the THEMIS Cydonia image, it would be an easy process to wait for his static IP address to log in to the official THEMIS website. At that point, Gorelick/Bamf who ran the THEMIS website could then redirect him, and *only* him, to a duplicate website that looked exactly like the official THEMIS page, but which instead contained the "real" data Keith had processed on his computer.

My confidence in this assertion is based primarily on the fact that this exact process—the act of redirecting someone to a counterfeit server or webpage, with or without their knowledge—has a specific name amongst the world's computer geeks. In the jargon of the techno-savvy, of which Noel Gorelick is undeniably a world-class member, it's called, simply…

Bamfing.

So what Gorelick/Bamf was doing all along in his months-long stay on Hoagland's forum was apparently seeking out just the right person to leak this authentic infrared data to. He found it in Keith Laney.

There was a firestorm of attention when we finally revealed this whole story to the public at large. Stories ran on MSNBC's website authored by Alan Boyle and in several local newspapers in and around ASU. But the reaction we got from members of our own "team" was curious, to say the least.

Many of the other independent researchers in the Mars research community saw the images and promptly panicked. After spending years debating the artificiality of Cydonia, when confronted with proof of it in the form of Keith's images, they reacted angrily. We were surprised to read in various forums that many of these supposedly curious souls were accusing Hoagland and Laney of having faked the "real" data. They were apparently more inclined to believe this than to believe that we had actually been leaked data that proved Cydonia was artificial.

Unfortunately, one of the most strident members of the board was Holger Isenberg. Once he heard that there were two sets of data (and we released them side by side on our site), he ceased all attempts to process the data. No matter how hard they tried, neither Keith nor Hoagland could get him to simply take Keith's version of the raw data and process it as he had done with the "official" version. After we posted our analysis of the data and laid out the events surrounding Keith's acquisition of it, Holger then turned on us and accused Keith of creating "fake" images and Hoagland of "lying" about how Keith had obtained the raw data. He then went on to declare that Gorelick was "his friend," and that his newfound friend (whom he'd never even met face to 020face) would never do anything like post degraded data in place of the real stuff on the ASU-THEMIS website, even though Gorelick had publically declared his ability to do exactly that. Finally, he alleged that the "real" data was created by scanning a print copy of some "fake" images created in Photoshop (an idea we later learned he was fed by Gorelick in a private chat on Hoagland's forum). When it was pointed out to Holger that *he* was the one who verified that the header data on Keith's file confirmed it came from a well-established NASA file format and was processed on a UNIX machine (there is no Photoshop for UNIX, at least at that time), he broke off all contact.

We found this process exasperating, but sadly confirmatory of the "Brookings" model, which said that the most technically accomplished amongst us (the scientists and engineers) would have the hardest time dealing with proof of a technologically superior civilization. Ironically, "Cydonia" derives from a Greek root word meaning "enlightenment"—but the evidence we presented only seemed to create fear and confusion, even amongst those who feigned interest in getting to the truth. Even friends like then *Coast to Coast AM* host Art Bell were not immune.

Art was having a hard time grasping the idea that the images Keith worked on couldn't be faked. He was convinced that Gorelick or Christensen had somehow embedded the images of various buildings and structures in the data. But, the meticulous process of creating thermal IR images prevented any such fraud.

Keith was essentially working with nine different individual images, each slightly different from the next due to the fact that they each came from a different IR filter wavelength, and from the motion of the spacecraft. Keith then cropped the nine images out, rotated them to vertical, and hand-aligned and assembled them into single frames, one layer on top of the other, two at a time. In other words, he took image strips one and two of the nine and laid one on top of the other, and then he took three and four and laid one on top of the other, and so on. This makes eight possible layer combinations for every individual image strip, adding up to a total of 72different possible wavelength/filter combinations. Next, he "color-ratioed" the combined images, assigning a color value (and weight) to individual elements in the image. This means that, as an example, titanium will appear "red" in the given image, if he assigns that color to that material. However, without calibration, i.e. knowing what the materials are in a given image, all you can do in actuality is assign *a* color value to *a* material. It tells you nothing about what that material is, just whether it is the same as other pixels around it. The same rules apply to heat differences in a given image. You won't see a temperature, but you will see a temperature differentiation.

Next, Keith used a decorrelation-stretch tool on the images. This is a critical tool that exaggerates (not creates) the color/material/ heat signature differences in the color-ratioed images. Then, as a

Coherent buildings and structures the size of city blocks on the THEMIS infrared data.

last step, he added the luminance layer of the visual image.

In short, there is no way that Christensen or anybody else could have possibly painted "fake" buildings on each of the nine images and guess how he'd combine each layer, what color ratios he'd use, what settings he'd use in the decorrelation-stretch tool, how he would rotate, align and crop the images, and whether or not he'd then combine the visual image on top of that. It's just flatly impossible. So the objects we were seeing in Keith's processing were definitely "there" on the real IR image.

Next, Keith endured a series of attacks from various amateur anomaly hunters who claimed it was he that "created" the blocks and buildings on the images. Further, they argued that nobody knew for sure that Bamf was really Noel Gorelick (despite Bamf's encyclopedic knowledge of infrared image processing) and that he might have led Keith into creating imaging artifacts with the imaging suggestions he'd made.

At this point, we had had enough. It was arranged for Keith to brief Art in more detail, and show him a comparison between the earlier *Phobos II* IR image and the new Cydonia data. That did the trick with Art on the images, but everybody still wanted proof of who "Bamf" really was.

So I called him.

It took no time at all to find Noel Gorelick in the ASU phone directory. I called him the afternoon of September 6, 2002. In the course of the conversation, Gorelick freely admitted he was "Bamf," that he was responsible for all of the postings on Hoagland's message board under that name, and he reiterated his stance that the Cydonia IR image page had been untouched since he posted it on July 24, 2002. I then went on *Coast to Coast AM* that night to pass this information on to the public at large.

But at that point, we had reached a stalemate. We had proven, through highly technical arguments not easily understood even by the technically adept, that Keith Laney's version of the Cydonia data was real, and the image currently on the ASU website was a deliberately degraded copy made from it. However, Christensen and Gorelick were intractable in their insistence that nothing had been changed on the ASU website where the data had been obtained since the moment it was first posted. We had many critics beyond the usual suspects like Mac Tonnies—a self-appointed Cydonia "expert" who regularly attacked everything we published—who insisted that "we" (meaning Hoagland, Laney or myself) had somehow "faked" the original data. Even some of our supposed allies decided to take that stance rather than admit that NASA or its affiliates would be involved in an outright scientific fraud (see chapter 4). I guess they had conveniently forgotten about the "Catbox" and MOLA frauds that NASA had previously perpetrated.[1]

But sometimes, if you are just a bit patient, life throws you an opportunity. Sometimes your opponent makes a mistake. A really big mistake.

We had discovered that both Michael Malin and Phil Christensen would be attending a public event at JPL's 6th international Mars Conference in Pasadena.[2] We felt like this would give us a great opportunity to confront Christensen about the clumsy attempt to discredit us the previous year. So on July 23, 2003, I dutifully attended the public event. Both Dr. Malin and Dr. Christensen gave slideshow presentations. Christensen could not resist showing the more recent five-band color image of the Face on Mars. As he brought the image of the Face up on the screen, he smirked and asked the audience if they knew what it was. There was a nervous chuckle throughout the crowd. He then put forth a new idea about the natural geologic evolution of the Face. He suggested that the unusually bright reflectivity of the eastern side of the Face mesa was because of an accumulation of carbon dioxide snow on that side of the Face. He smirked once again when he asked the audience if they had any comments on his idea (there were none). However, a few moments later, when Dr. Arden Albee opened the forum for public questions, the smirk was wiped off his face.

As no one stepped to the microphone initially, I decided to spring my trap. Dr. Albee suggested that the person asking the question give their name. When I approached the microphone and gave my name: "Michael Bara," Dr. Malin audibly groaned and proceeded to hide behind a stage curtain. Malin emerged from behind the curtain only after hearing that my question was for Dr. Christensen.

Christensen himself became highly agitated. He began pacing back and forth across the stage as I asked him if the original Cydonia daytime infrared image, which was posted on the ASU website on July 24, 2002 had ever been changed. At one point as he stumbled for an answer, Christensen dropped the battery pack for his wireless microphone and had to scramble to pick it up. Obviously, he was surprised and caught off guard by this confrontation. When he did answer, it was in a nervous and halting tone, and he only occasionally made eye contact with me.

Christensen defended the data on the THEMIS website. He claimed that once it had been posted on July 24, 2002 it had never been altered. He also went on to state that he had no idea how the "artificial stuff" as he called it, had gotten on the images that Keith Laney had processed. At no time did he say that the Laney version of the image was an outright fake, nor did he accuse Hoagland, Laney or me of creating false data. His choice of the words "artificial stuff" was quite telling, because he could have just as easily used the word "fake." He pointedly did not.

Just to be sure, I asked for clarification on the most crucial point: had the data that was posted on the ASU/THEMIS website

Screen capture from the Internet Archive showing that the THEMIS website containing the Cydonia IR images was altered on July 26th, 2002, directly contradicting the statements of ASU's Dr. Philip Christensen

been changed after July 24, 2002? Christensen again emphatically answered "no." What he did not know at that time was that at precisely that moment, we had him.

Unbeknownst to Christensen, there is a website called the "Internet Archive" (archive.org). Sponsored by the Library of Congress, its purpose is to document every page that has ever existed on the Internet. Using a search engine called the "Wayback Machine," I had been able to plug in the URL of the daytime Cydonia infrared image. It had shown that, contrary to Christensen's previous e-mail statements and contrary to this very public statement, the ASU website containing the image *had* been altered. Created and managed by Christensen's lieutenant Noel Gorelick, an image had been originally posted on July 24, 2002. This is the one Keith Laney (and perhaps a handful of others) had downloaded. But, completely at odds with all his statements, the archive showed it had subsequently been altered in the early morning hours of July 26, 2002. As you'll recall from the previous chapter, Laney downloaded the "real" version of the Cydonia daytime IR late in the evening (10:27 p.m.) on the *25th*. In other words, just hours after Keith Laney had downloaded the "real data" from the THEMIS website, Gorelick, Christensen or someone else had changed the contents of the site and updated the page.

So, all of Christensen's protestations to the contrary, there had been at least one change to the Cydonia infrared image website *after* Keith Laney had downloaded the "real" data. Within a few days of me publically confronting Christensen with this revelation, the Internet Archive ceased tracking changes to any of the THEMIS/ASU image release pages, and deleted the records of the controversial page itself. This can only be done at the request of the website owner, who in this case would have been Noel Gorelick. So it was clear we would not be trapping Christensen or Gorelick in any more lies anytime soon.

It was hard to tell from Christensen's reaction whether he was so nervous because he was afraid he was going to get caught in a lie by me, or whether he was under some pressure from outside forces to keep the story straight even after this encounter. I'm still not certain whether Christensen was a friend or an enemy of the independent investigations. On the one hand he had given us the data to prove our thesis beyond any reasonable doubt. On the other hand, he had participated (at least tacitly) in an attempt to discredit the investigation publicly by spreading false information and data to the public. Yet still, he had gone to some lengths at the conference to avoid accusing me or anyone else of faking the "real" IR data. In the end, I simply didn't have enough information to reach a conclusion about Christensen the man.

But the confrontation seemed to send shockwaves through the planetary science community, which promptly stepped up its attacks against Richard Hoagland, then the public face of the investigation. This came in the form of a newly minted debunker, Phil "Dr. Phil" Plait, a grotesque little toad of man who fancies himself the successor to Carl Sagan as the debunker of all things paranormal or supernatural. Following in the footsteps of such unpleasant characters as James Randi, Jim Oberg and Phillip J. Klass, Plait created a blog site he called "Bad Astronomy," which he used to attack (in the most personal manner possible) anybody who dared challenge the established NASA orthodoxy on UFOs, Mars or even climate science. Of course, he, like so many before him, takes pains on his website to stress that he does not and never has "worked" for NASA—he just takes grant money from them.

Plait entered into the Cydonia fray after my confrontation with Christensen at Cal-Tech, dedicating a whole series of pages to attacking Hoagland specifically. Most of the information is just a rehash of shopworn, discredited arguments I covered in *Ancient Aliens on Mars*, including the now infamous "Old Man in the Mountain" nonsequitur. In fact, Plait seems genuinely obsessed with Hoagland, and frankly jealous of the amount of attention he is able to generate. This leads him to make various bad arguments, false claims and even to perpetrate outright lies. But by far, the biggest lie is the one he asserts on his page about the Cydonia IR data.[3]

He starts with a bold, personal and outright false statement about Hoagland:

Hoagland's stock in trade is in grossly misinterpreting images, usually by taking compressed JPEGs and then enlarging them beyond their resolution. Sometimes he just looks at images and, it seems, makes stuff up out of thin air. Hoagland wrote a whole book on the "Face" on Mars, when it is really just a hill. But he does this over and over again, with many such images.

This is first off just plainly false. Neither Hoagland's book, *The Monuments of Mars,* nor our collaboration *Dark Mission,* constituted a "whole book" on the Face on Mars, any more than *Ancient Aliens on Mars* did. Hoagland also very rarely processes the images he discusses, although after 35 years of working with digital images, he's pretty good at it. Whenever possible, the images used are not "compressed JPEGs" but full resolution source images, usually lossless file formats like TIFFs. They are simply copied to the JPEG format after processing for placement on the internet at a reasonable file size. He also never "enlarges them beyond their resolution." This is just another false claim constantly spread by critics of the Cydonia investigation ever since Carl Sagan made the assertion in a *Parade* magazine hit piece in 1985. But what Plait deliberately hides from his readers is that the majority of the image enhancement work done over the years on the Face, Cydonia and other Martian anomalies is not done by Hoagland at all, but by contractors, scientists and interested professionals with vast years of experience. Dr. Mark Carlotto, Vince DiPietro, Greg Mollenaar and Keith Laney actually do imaging work for NASA, not Hoagland. But somehow Plait just failed to mention that.

He then goes on to push the JPEG argument even further (I apologize in advance for forcing my readers to look at pictures of Dr. Plait):

Take a look at the images below. The image on the left is a JPEG of me saved at medium compression. Click on it to get the full-resolution image if you want to see the nicest quality (it's about 70kb). The image on the right was saved with maximum compression. The image looks awful. Straight lines in the higher-res image look

Phil "Dr. Phil" Plait

crooked, or wavy. Curves become blocky; colors get oddly mixed."

To make things worse, literally, I cropped the original high-resolution image to just show my right eye, resized it to twice its original size, then saved it at maximum compression. Look how awful the quality is. You can see all sorts of wiggles and weird features around my glasses. None of that is real, as you can see in the high-resolution image. They were all introduced when I compressed the image.

He then finishes his argument off by claiming that artifacts in the Cydonia IR images (which he links to) are the result of the same kind of "image compression:"

Hoagland's website is full of images like this. He takes images, blows them up, saves them as JPEGs, then claims there are patterns in them indicating regular structures! Here is one he claims shows a building. Here is another with a "power plant". In fact, this long, rambling webpage on Hoagland's site has many such examples of over-magnified, over-compressed images showing compression noise. They are not artificial structures, as he claims, they are simply what you get when you over-manipulate an image, as I did above.

Now for reference, the "building" he is referring to is the "Temple" from the THEMIS infrared images, and the "power plant" is another object that I identified in the article about them. All three objects are anything but compression artifacts, as Plait claims, and this can be proven. It will also be shown that Plait was well aware of this when he made his false claim, and that he did this as a deliberate deception upon his readers.

The "Temple, the "Train Station" and the "Power Plant" from the Cydonia THEMIS IR data. All three coherent structures are far above the threshold of compression artifacts.

First of all, no matter what Plait says, compression artifacts simply do not create false structures on data. Certainly they do not create objects that show identifiable signs of architectural design intent, like tube-like objects that terminate at the base of buildings with openings in them, as an example. Further, JPEG compression artifacts always run north/south in an image. The block-like city structures do not. They run true *Cydonia* north/south, and are present weather the image is rotated or not. This means that they could not have been introduced when the image was rotated.

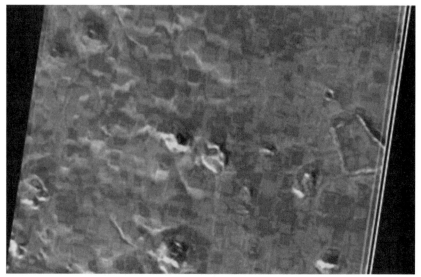

THEMIS IR image showing block structures running true Cydonia north/south, rather than image scan north/south. This is proof that block-like structures are not compression artifacts, as alleged by "Dr. Phil" Plait.

In addition, compression artifacts are well understood and easily accounted for by imaging specialists like Keith Laney and Dr. Mark Carlotto, who certainly would never mistake them for real buildings on (or under) the Martian surface. This is why Plait wants his readers to believe that Hoagland did all the enhancement work himself, and by implication either doesn't understand what JPEG artifacts are (despite having discussed them *ad nauseum* in various articles and books over the years) or worse, is intentionally deceiving his readers.

The reality is quite the opposite, in fact. It is Plait who

is deliberately deceiving his readers in an attempt to discredit Hoagland.

Somehow, Phil "Dr. Phil" Plait managed to miss one very important fact that blows his "JPEG compression artifacts" thesis out of the water—none of the Cydonia thermal IR images processed by Laney, Hoagland or myself were JPEGs. They were, in fact, all TIFF files, a compression-free image format that makes it impossible to introduce such artifacts. In fact, had he actually read the story he linked to (which he asserts he did, since he refers to it as "rambling"), he would have noted that it included links to the full size TIFF files so others could also work with the data. The JPEG files he refers to were simply browse versions for quicker download times.

How such a genius astronomer, one who thinks enough of himself to spend so much energy criticizing the research and personal integrity of others, could have missed this point is a mystery. Unless he knew it all along. Unless he knew full well that all the processing work had been done on full resolution, uncompressed TIFFs which don't contain any compression artifacts. Unless his clear intent was to deceive his readers from the get-go.

It was.

So while Plait and Christensen's mendacities were fairly easily dismissed, there was still the matter of the guys on our own side we had to deal with. And the possibility of getting something we wanted even more than a daytime infrared image of Cydonia: one taken *at night*…

(Endnotes)

1 Ancient Aliens on Mars, the author, chapter 7, 8

2 http://www.caltech.edu/content/sixth-international-mars-conference-will-include-public-event

3 http://www.badastronomy.com/bad/misc/hoagland/artifacts.html

Chapter 4
Night, and the City...

The attacks on the Cydonia infrared data we had presented were actually very predictable, especially coming from the likes of "Dr. Phil" Plait. The THEMIS processed images were nothing less than the "smoking gun" data the independent researchers had been seeking for decades. There was never any question that those who had staked their professional reputations and their continuing grant money on the NASA "it's just a hill" orthodoxy would not only misrepresent the data, as they had done so many times before, but vilify us as heretics. What we all should in retrospect have known, though was that there would be those in the independent research community who might take the same stance. Just as the Brookings report had predicted, the revelation of the truth of a superior Ancient Alien civilization would be too much for some to bear. Holger Isenberg's reaction was a prime example, as was the reaction of a former ally, Dr. Mark Carlotto. But Carlotto's was certainly the more disappointing.

Shortly before we published our revelation of the existence of the two THEMIS Cydonia IR images, one real, and one a highly degraded scientific fraud, Mark Carlotto put out his own analysis of the Cydonia multispectral images. From the beginning, we were concerned about the contents of the Carlotto analysis. For one thing, he made a number of simple errors. In one case, the IR image is referred to as "E0201847.gif," which is the wrong file name for the multispectral Cydonia image, "20020724A." In addition, there were numerous misspellings and other obvious mistakes which gave the whole project an air of haste and sloppiness. It did not seem to be up to Carlotto's usual thorough standards—at least as Hoagland had remembered them from working with him years before.

As we got into the content of the article, it seemed to stray into even odder territory. Carlotto started out by comparing the *Odyssey* Martian THEMIS data to terrestrial Landsat images, a very inaccurate comparison to say the least. Landsat is a 1970s-era technology that produces primarily surface reflectance data in the visible region of the spectrum, as compared to THEMIS, which concentrates on extracting data primarily from thermal infrared emissions from surfaces and objects. Landsat, in contrast to THEMIS, has virtually no ground penetration capabilities at all. A far better comparison would have been made to a relatively new Earth orbiting instrument (1999) called ASTER (Advanced Spaceborne Thermal Emission and Reflection), which has very similar near-infrared capabilities to THEMIS.

Carlotto went on to make some even odder errors in his article. He proceeded to conclude that there were various clays making up the composition of the region surrounding the Face. However, Carlotto surely must have known that he could make no such conclusions, since the calibration data for either the official version or the real version of the Cydonia IR data had yet to be released. He also failed to perform a decorrelation-stretch, a step that is crucial for separating the thermal data from the composition information in the image. However, even if he had done the decorrelation -stretch, it would have made no difference, since without the crucial calibration data his conclusions about the specific composition of Cydonia were meaningless. Beyond that, he completely ignored the overwhelming amount of noise in the official image and seemed more than satisfied with the poor quality of it.

Even as we tried to make sense of Carlotto's seeming brain fade, our attention was drawn to another recent item posted on his website. In it, he directly addressed the issue we had raised concerning the discrepancy between the two IR datasets, the "official" one as posted by ASU and the "real" one obtained by Keith Laney. Carlotto declared that clearly the "real" version we had posted was a degraded version of the "official" one, exactly the opposite of what the data indicated.

We knew this of course to be utterly laughable. In his analysis of the two images, Carlotto had taken only a single unnamed band from the "real" data, and compared it to a single unnamed band from the "official" version. He had not done a full composite of the "real" image bands, had not done any color ratios, nor performed the decorrelation-stretch to enhance the data. All he did was make a simple visual inspection of two grayscale bands without performing any of the accepted processes for enhancing false-color thermal IR data.

This is flatly not how you handle thermal infrared data. This is higher-order IR information, compared to a simple pretty picture. Even though it has a lower *visual* resolution, it contains information totally unavailable from even much higher resolution visual images. That is what makes lower resolution thermal IR THEMIS data such a potential breakthrough on the 30-year-plus Cydonia problem, even at around 100 meters per pixel. It is this fundamental fact of infrared optical physics that made us believe from the beginning that Laney's July 25th Cydonia IR image was the real deal, and the later apparently much sharper (but noisier) official image was, in fact, the doctored version. Quite simply, Laney's version contained more data.

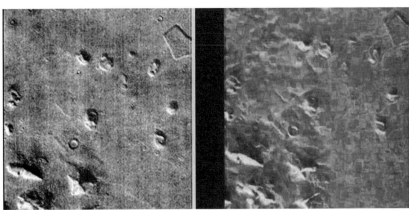

Doctored "official" version of Cydonia THEMIS IR data (left) and much cleaner "real" version (right). Histograms show that the "real" version contains far more optical data than the "official" one.

The "real" image, while it may look a bit blurry, actually contains far more data than the "official" version, no matter what the "official" version looks like on first (naked eye) visual inspection.

That's why the processing and enhancement tools used for multi-spectral imaging are crucial to extracting that "hidden" signal from this set of IR images.

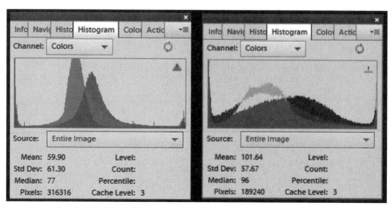

Histograms for the "official" THEMIS image (left) and the "real" Keith Laney processed image (right). The "real" image contains far more data and dynamic range than "official" version, a tell-tale signature that the "official" image is an altered and degraded version of the "real" one.

Carlotto, it seemed, was making the same elementary error that some of our readers were making—assuming that sharper is automatically better, as it is for *visible* light images. And curiously, he did not do the other crucial steps which would have instantly shown him that the "real" images contained far more (and far better quality) data than the official version did. This simple analysis in and of itself would have immediately disproved his core hypothesis. For, how can "degraded" data produce quantitatively better results under proper analysis than its supposed source data, and with infinitely less noise? The simple answer is that it can't—but that's not what really bothered Hoagland and me. What troubled us in the extreme was that Dr. Carlotto, a world-class imaging expert, and a DOD contractor on a host of classified imaging analyses, *should* have known all this.

So, given this decidedly odd state of affairs—a well-known and respected imaging specialist, who at least *used* to be curious about Cydonia and suspicious of NASA, making not only a seemingly colossal error in judgment, but compounding that error by failing to simply put it to the test—we decided to consider our options.

Hoagland, Keith Laney and I held a conference call about Carlotto's article on the night of September 3. The general consensus was that Carlotto had effectively "polished a turd" and declared that he discovered a pearl—all without even considering the field of gems which had been placed right in front of him.

Hoagland, however, refused to buy into the notion that Carlotto was as incompetent as his analysis made it seem. He staunchly defended Carlotto's skills and professionalism, insisting that there must be some other reason for his reticence to properly process the THEMIS images. But, in the interim, Carlotto had fallen in with a group calling itself the Society for Planetary SETI Research, or SPSR. SPSR was an organization of independent Mars researchers with mostly academic backgrounds who staunchly refused to believe that NASA would ever lie to anyone about anything. We agreed that it was possible that Carlotto had simply fallen in with the "honest but stupid" crowd that forgave every NASA transgression, yet held us to much higher standards. The decision was taken to reach out to Carlotto—if only to save him the embarrassment of a public response pointing out his lack of thoroughness. I subsequently e-mailed Carlotto, informing him he'd made several errors in his article, and advising him to pull it, at least until we published our own analysis.

Carlotto responded via e-mail and pointed out that he had plenty of experience working with thermal images, that the "real" image was "obviously" degraded, and that his paper had been peer reviewed (by Dr. Horace Crater, an SPSR statistical analyst who had no working experience with thermal infrared data). My response back was basically "suit yourself, but if we had peer reviewed your paper, it would not be published right now." Carlotto then e-mailed me back, got Hoagland's phone number, and the two men had a chat on September 4, 2002.

According to Hoagland, what Carlotto seemed most concerned about was that his previously published paper would be made obsolete by our article. After a wide-ranging discussion, which included Carlotto pointing out that he'd written his own decorrelation-stretch algorithm (even though he didn't use it on the

THEMIS data), Carlotto agreed to take the "real" image, perform all of the proper steps (composite, color ratios and decorrelation-stretch) on it, and either call or e-mail Hoagland with his results. That never happened.

I have no way of knowing if Carlotto ever did the analysis he agreed to do, but after more than a week of waiting, Carlotto's only response was to publish yet another update on his web site in which he dug himself into an even deeper scientific and ethical hole.

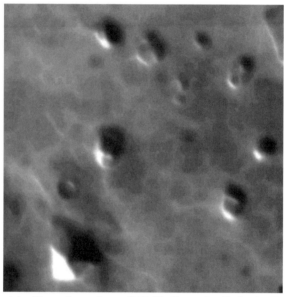

Deliberately blurred version of Keith Laney's Cydonia IR image created by Dr. Mark Carlotto.

Instead of following the proper protocols for processing thermal infrared data as he'd agreed to do, Carlotto decided instead to take the "official" version of the Cydonia THEMIS image and subject it to a series of contrast and blur filters in an apparent attempt to "prove" that the Laney image was generated by degrading the official one. He did this by taking only a single band image, not a composite, and he of course did not do any of the other tests he agreed to perform in his conversation with Hoagland. This led him to conclude "its similarity to the top left ["real"] image strongly suggests that the *Enterprise* (Laney) image is an altered version of the ASU image."

This assertion is easily disproven by simply looking at the histograms from both images.

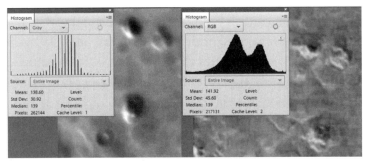

Carlotto's altered "official" Cydonia IR image and Laney's "real" version, right.

Carlotto's altered image contains only a fraction of the data that Laney's image does. This is because the image he started with, the "official" ASU fraud, *already* contained less data when he started. Further, the contrast and blurring he deliberately introduced into the image in an attempt to make it look like Laney's reduced the image's data content even further. As the histograms show, Carlotto's version contains only a few scattered bits of data (actually shades of color) separated by wide bands of no data at all, while Laney's shows a broad band of shading across the full 256 shades of gray/color. If Laney's image was made from the "official" ASU version of the IR data as Carlotto flatly claimed, the histogram for his version would look almost exactly the same as Carlotto's. The fact that Carlotto could not reproduce what Keith had done without drastically reducing the data content is proof that Carlotto's thesis is

Animated GIF frames produced by Dr. Mark Carlotto showing subtle shifts in subsurface structures on Keith Laney's version of the ASU Cydonia IR data.

wrong. Personally, I would have thought that an imaging expert like Carlotto might have checked and compared something as basic as the histograms himself before publishing an article that made him look so stupid, but hey, that's just me.

He then doubled down on stupid by claiming that because the Laney image changes from band to band, it is a "distorted" version of the "official" ASU release. What is truly disturbing is that Carlotto's "test" here is proving exactly the *opposite* of what he is claiming. Real multispectral data (and certainly thermal IR data) *does* change from band to band. What he is illustrating is exactly what true multispectral data should look like. He's not seeing *distortion*, but rather the expected shift in "return" to the camera from slight variations in the thermal signatures of actual features on (and under) the Cydonia plain. The other phenomenon he seems to be describing, the "shift" in certain edges of some of the large features, is simply due to the fact that the various bands are not all taken at exactly the same time. There are significant shifts in the spacecraft's position as different filtered CCDs in the THEMIS camera record (in a rapid-fire sequence) the actual imaging data for all the bands. This makes it a near impossibility to simply overlay the various bands when doing a composite. Of course, had he simply read the camera specs, he would have known this, and physically corrected, as Laney did, for the minute geometric shifts.

That Carlotto made no effort to correct the alignment problem is not only a testament to his lack of thoroughness in this case, but an outright indictment of his methods and possibly even his motivations. Keith Laney, at Hoagland's request, produced an image similar to the section that Carlotto had done, only with the various bands properly aligned geometrically. It took him all of five minutes.

All Carlotto would have to do, if he truly wanted to decide the question of which dataset is degraded and which is "pristine," would be to run the two images side by side through a quality enhancement tool as Keith had done. Had he done so—as he promised Hoagland he would—he would have clearly seen that he got it completely backwards.

What makes this truly egregious is that Carlotto certainly knows everything I have described above, that a simple visual inspection of a single grayscale IR band is not a valid comparison of these incredibly information-rich datasets. He evidently decided that it was better to try to cover his own mistakes by making up a pretty picture which only proved his talents as an artist, not as a scientist. What seems to have happened is that Carlotto was unwilling to publicly face the fact that his initial declaration that the Laney image is the degraded one is flat out wrong. Given the opportunity to admit his mistake, he instead decided to cover his tracks with this absurd comparison. We were truly sorry that Carlotto had chosen to take a politically defensive stand, instead of the scientifically courageous one.

The absurdity of his position was underscored by an e-mail Laney received from Research Systems, Inc., the creators of the ENVI 3.5 software that he was using to process the images. While refusing to get into the middle of the controversy over which image contained the original source data, the communication made several points totally inconsistent with Carlotto's analysis:

I must admit, this has made quite a stir in the astronomy community! At any rate the images look awesome! As I tell anyone who asks about RSI's stance... 'RSI does not have an opinion either way, we just want to provide the best software to scientists so that they can do their own best work.'

How could someone crudely degrading the official Cydonia data create "quite a stir in the astronomy community?" Given the scientists that RSI routinely deals with, and their level of multi-spectral expertise, any simple degradation of the official ASU THEMIS website image into the one Laney had been working with (with RSI looking on) would most certainly have been caught if it was as simple as Carlotto was now claiming. Also, if Keith's image was a hoax, why would the RSI representative go on record saying that the images produced from it "look awesome?" Wouldn't the better part of valor be to simply refrain from all comment until the lineage of the "real" image was determined?

In effect, Carlotto had simply parroted the position of Dr. Philip Christensen of ASU by declaring that the Laney data is "degraded," when all parties involved (as the RSI e-mail underscores) certainly *know* that the official version is far inferior to the Laney data. By refusing to put that data to the true scientific test—whatever the reason—Carlotto and his SPSR colleagues were reduced to being nothing but mouthpieces for the NASA party line. As I said before we released this data, people were going to have to take sides... and SPSR and Carlotto evidently had.

However, the attempt to bring Carlotto back into the fold was not all wasted. A reader, Wil Faust, made a truly inspired suggestion

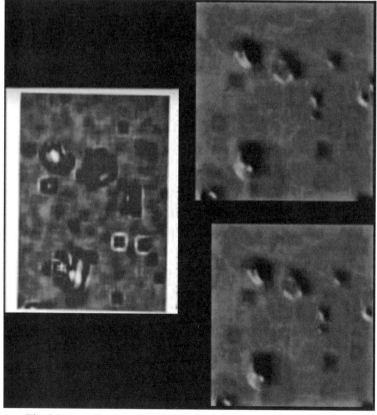

From *The Martian Enigmas: A Closer Look* by Dr. Mark Carlotto (left); and from Keith Laney's "real" version of THEMIS IR data. Laney image enhanced by contrast stretch only.

to us. Why not compare, he said, Carlotto's own seminal work—his fractal analysis of the Cydonia region from the *Viking* data (see *Ancient Aliens on Mars*)—to our own IR results?

So we did.

It turns out that when you use Carlotto's own methodology and take a single band of the IR image strip from Keith's "real" version of the data to compare it against Carlotto's own fractal analysis of *Viking* frames 35A72 and 70A13, you get quite striking results. Not only are the THEMIS blocks—which Carlotto flatly claimed were enhancement artifacts—clearly visible on his own work from late 1980s, but they match up very *precisely* with the blocks on the Laney image, literarily one for one. What this means is that Dr. Carlotto's own process is identifying exactly the same faint, subsurface structures that Keith's enhancement showed. It is then incumbent upon Dr. Carlotto to demonstrate just how "filtering artifacts" can not only line up with features in *Odyssey's* THEMIS and visible light images—but also with direct non-fractal (non-natural) "hits" in his own dataset.

It is one thing to make honest errors in a piece of scientific work. It is quite another to compound those errors and miss the entire forest by hiding behind an incompetent peer review and obvious political propaganda, without even checking your own previously published work. Until he does this, I can no longer endorse his methods or his intellectual honesty on any issue pertaining to this continuing extraterrestrial artifacts investigation. Carlotto did some significant early work on the original *Viking* data, but when he made his ridiculous assertions about the THEMIS data he essentially played himself out of the game.

Fortunately, the world was about to get new data which had the potential to make all of the debate about disputed daytime thermal IR irrelevant—a nighttime thermal infrared image of Cydonia…

From the beginning what we (the independent researchers) really wanted was not the daytime thermal infrared data that Christensen and Gorelick had delivered, but a *nighttime* IR THEMIS image of Cydonia. The reason for this is simple and twofold. First, it is probable that any non-natural, artificial constructs in Cydonia or elsewhere would be made up of materials like steel, aluminum, titanium or glass, that were *different* from the background "natural" hills and mesas. They would therefore have very different thermal infrared signatures from their surroundings. They would likely absorb and retain heat much differently than simple rocks, dust or clays. This effect would be greatly heightened at night, when daytime solar radiation (heat) would be completely absent. This would create a higher "signal-to-noise ratio" in a dark, nighttime IR image, allowing easier detection of potentially artificial materials or formations.

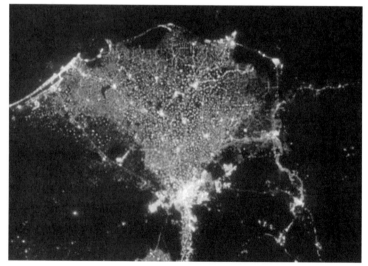

Nighttime thermal infrared image the city of Cairo (—al Qahira in Arabic meaning Mars) the Nile Delta and the Mediterranean Sea.

This concept can be proven out by examining thermal IR images of Earth taken from space. In the infrared band of the light spectrum, the inhabited parts of Egypt glow brightly. There can be no doubt from even a casual examination of this image that we are looking down at an intelligently-designed complex here on Earth. We don't even need to know what is *causing* the bright, glowing heat sources. It's their organized, geometric layout that is the tell-tale signature of intelligence, whether it is here on Earth or on Mars.

We got a small taste of what this data might reveal in the pre-dawn, five-band color image of the Face on Mars taken in October 2002 (see chapter 8, *Ancient Aliens on Mars*). It had shown us that the right, or eastern side of the Face was composed of something that made it incredibly reflective—far more than it should have been if it were a simple mountain or mesa, as NASA had repeatedly alleged. Arranged in geometric panels, this reflective material had all the characteristics of manufactured metal plates.

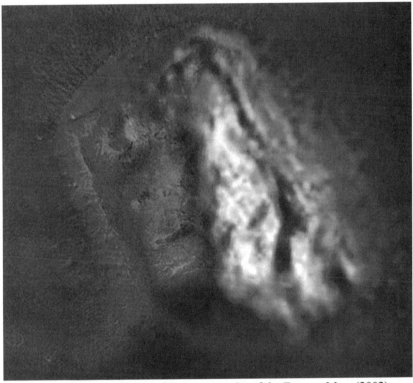

Five-band thermal IR/visual light composite of the Face on Mars (2002).

NASA's Dr. Philip Christensen had suggested at the Cal-Tech presentation that the right side was so bright in the early morning pre-dawn light because it was plastered with "pasted on" carbon dioxide snow. This conclusion was decisively refuted when MRO close-ups of the Face revealed no snow and distinct structural features.

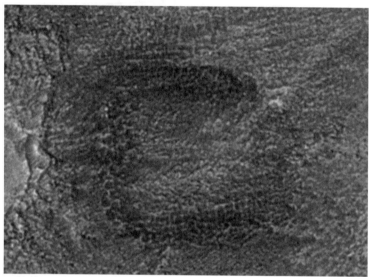

Collapsed structural framework on the right side of the Face on Mars.

Keith's enhancement work with the daytime thermal infrared images (remember, nine thermal IR bands, essentially nine different pictures of Cydonia) had confirmed that the "left" and "right" sides of the Face were compositionally different. A closer look at the same data not only confirms the geometric panels on the right side, but also suggests that the right side is more eroded and the understructure is more exposed. In other words, it has less casing, most likely because of wind erosion.

So it was with anxious if not breathless anticipation that we were all waiting for a promised eventual release of a nighttime thermal infrared image of the Cydonia ruins and the Face. It certainly had the potential to show the cityscape beneath the ice and the ruins themselves in levels of detail that would end the debate once and for all—provided we got the "real" data and it was the full nine bands like we had received from the daytime data.

Of course, we didn't.

When the much sought after nighttime image was released (on October 31, 2002, Halloween no less), all we got was a low resolution day/night side by side comparison. A poster, in essence,

Of course, they couldn't give us the real stuff. That would have ended not only the debate once and for all, but possibly ended

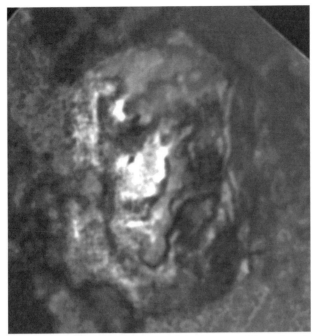

THEMIS daytime thermal infrared image of the Face on Mars showing distinct structural and compositional differences between the left and right sides.

several careers as well.

It was immediately clear that there were issues with the data. Rather than the full nine bands as we had been given in the daytime image of July 24–25, 2002, this time we got a cropped image showing

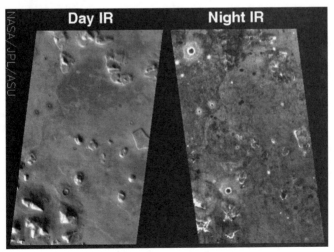

Trick… or treat?

a portion of a single nighttime image strip alongside a similar crop from the July daytime image. The official ASU caption read:

> *This pair of THEMIS infrared images shows the so-called 'Face on Mars' landform viewed during both the day and night. The nighttime THEMIS IR image was acquired on October 24, 2002; the daytime image was originally released on July 24, 2002. Both images are of THEMIS's ninth IR band (12.57 microns), and they have been geometrically projected for image registration.*

It was fairly quickly established, by looking at sun angles, heating patterns and noise levels that the new "nighttime" image was not taken when ASU claimed, on October 24, 2002, but rather much earlier in the Martian year, around January 2002, the dead of Martian winter. We also did not get the nine bands of infrared data we were seeking, but as I stated just the single IR band in a "poster" format.

Because there was so much digital "noise" in the image and we only had the single band, it was impossible to confirm the underlying block patterns seen in Keith's version of daytime IR data. There was even evidence that some of this noise was actually deliberately introduced into the image, to make such confirmation all but impossible. None of this prevented us from working with what we had, though.

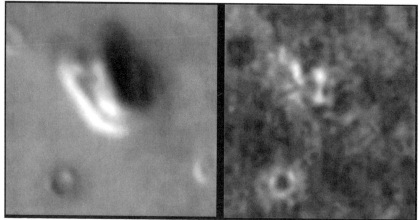

The Face on Mars from daytime thermal infrared (left) and nighttime (right).

The most striking and immediate thing we noticed about the nighttime image was something that was almost completely unprecedented—the virtual disappearance of the Face itself. In a side-by-side presentation, the Face as seen in the July 24th daytime IR image, compared with its nighttime October 31st counterpart— is essentially completely missing. The question is, why?

If the Face was just a perfectly normal natural formation, or "mesa," as NASA likes to refer to it, then it should cool at the same rate as all the other "natural" formations in the IR data. It doesn't.

As we looked at this sad little image—remember, just a tiny fraction of the data the *Odyssey* spacecraft surely acquired whenever the image was taken—it quickly became obvious that the Face is nothing like any of the other formations in the image. Other objects, like a clearly natural "slab mesa" to the right, the D&M to the south and a symmetrical mesa just south of the Face, do show up and appear clearly in the image.

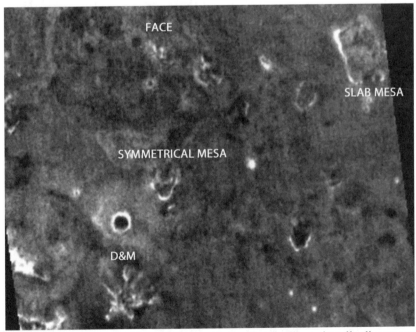

Portion of Cydonia nighttime IR image. Notice that the Face virtually disappears compared to the other features.

What this means is that the Face cools at a different—and much more rapid—rate than the surrounding structures. What then follows is that it is made of much different material than these other "mesas." As we already know from the five-band pre-dawn color image and daytime thermal infrared images, the Face has highly unusual properties for a "natural" mesa. In fact, it is nothing *like* the surrounding formations. Based on their cooling rates, these other objects are most likely made of some kind of stone. Stone cools much more slowly and retains heat better than other materials, like metal for instance. So even though the Face effectively does not appear in the nighttime IR data, that alone tells us two things; it is different from any of its nearby allegedly "natural" formations, and it's made of something that cools much more rapidly than they do. Most likely metals of various kinds. And that makes it most probably artificial.

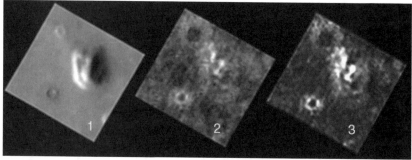

Daytime IR image of the Face on Mars (1), nighttime (2) and averaged (3).

Because ASU had at least put individual IR band images side by side on their poster release, it made it easy for us to take the two images and overlay them one on top of the other, essentially averaging the two datasets. Keith Laney quickly did this, and when he did, miraculously the Face reappeared. When it did, what we saw was that there was a box-like, very geometric underlying support structure to the Face, one which had not been revealed in visible light images or even the initial daytime thermal infrared images. Almost assuredly, this box-like structure is the underground foundation for the Face itself, only revealed because of the unique ground penetrating capabilities of the THEMIS instrument.

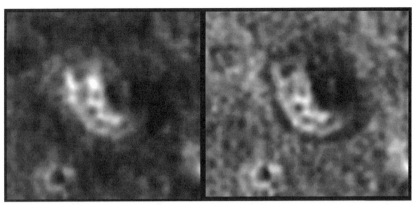

Two versions of the Face in an averaged daytime/nighttime thermal IR image.

But, doesn't this new information, as solidly reinforcing of the artificiality premise as it is, mean that other objects, like the D&M, are not artificial? After all, they seem to follow a more natural cooling curve, as if they are made of stone rather than the exotic metals the Face seems to be composed of.

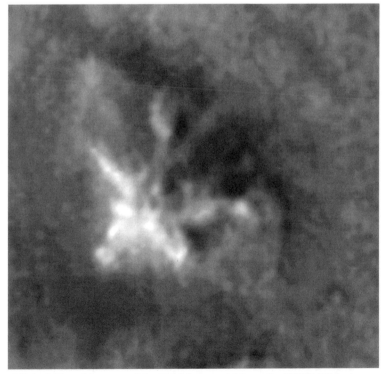

Day/Night infrared images of the D&M pyramid from the THEMIS instrument.

The truth is that no, that's not what it means at all. All it means is that in this particular dataset—which is only one leg of the artificiality hypothesis—the Face is more "unnatural" than the D&M. The D&M is still plenty weird.

Unlike many of the familiar surface features at Cydonia, which seem barely able to retain any daytime thermal energy, the D&M is almost incandescent in this nighttime view. In fact, it's lit up like a Christmas tree, almost as if it were internally still warm. This conclusion is supported by the fact that in this pattern of nighttime illumination, the thermally "hot" areas are not restricted just to the sunset side (left) of the D&M, but also in the shadowed right-hand side. This is different from the other natural formations visible in the nighttime infrared. Given the incredibly cold temperatures at night on Mars (-140C, at least), this is all but impossible... unless the D&M is made out of some kind of ultra-heat retentive material, or there is an artificial and still active internal heat source.

Like, a furnace of some kind.

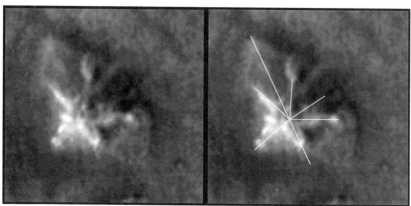

Geometric layout of the D&M pyramid in day/night infrared.

Also clearly evident in the new image is that the D&M is *structurally* inconsistent with any kind of natural object. Its pentagonal, multi-symmetrical layout seen in the visible light images is strongly reinforced here, and we even see a deeper underlying structure that is only evident in the infrared. Natural formations simply don't exhibit this kind of geometrically perfect substructure.

So in the end, the long sought after nighttime infrared image of

Cydonia only confirmed what we already knew about the region. It did not substantively expand it. Had NASA/ASU seen fit to provide the scientific community with a full set of the unmodified nine-band thermal IR data, there is no telling what it could have revealed. At the very least, we would most likely have had confirmation of the underground (or under-ice) city that the daytime image had exposed. At the most, we would have had the smoking gun we had been looking for. Instead, all we got was a small, incremental step forward in our attempts to verify the artificiality hypothesis at Cydonia.

As of this writing, more than a decade later, as far as I know none of the other nighttime IR data taken by NASA/ASU for the October 31, 2003 poster has been released. Gorelick himself admitted on Hoagland's message board that he had the data, but didn't think he should have to release it to the public which paid for it. *"I could do a 9-band IOTD like the 7/24 image, but since we've already done of it [sic], I don't think we'll do another one."* [1] When this comment was passed on to mainstream space news sources, like MarsNews.com, even they commented on the relative dearth of data being released by the THEMIS instrument.

Jim Burk, Editor-in-Chief, responded:

Oh, thanks for posting your chat with Gorelick. It is incredible to me that he feels no obligation to post the entire raw data set for the night-time image if he has it. If he had nothing to hide, he would have done that on Oct 24th (the day it was acquired.) I would even give him 24 hours for any conversions/etc. that needed to take place. I have never understood why they always take days, weeks, or months to give the public anything. I'm quite sure that what eventually (after MONTHS) shows up in the PDS is not the whole story...

"Months" has now become more than a decade, and still this data has not been released. If not, why not? Is it possible that it would reveal just what we have always said it would reveal?

Our only hope was that somebody else would send a probe

to Mars, and return honest data. Like maybe some clear color images of Cydonia. As it turned out, we'd have several years to wait, and NASA was far from done with their manipulations and misdirections...

(Endnotes)

1 http://www.enterprisemission.com/nighttime.html

Chapter 5
Mars Express

Even after the Cydonia infrared image debacles of 2002 and 2003, we (the independent researchers) still held out hope that we might get some good data from other sources. While the thermal IR data was valuable, by the mid-2000s we still had not gotten a single direct overhead visual image of the Face or a single color image of Cydonia. But a new probe from the European Space Agency (ESA) named *Mars Express* held the promise of delivering both.

Carrying an instrument called the HRSC (High Resolution Stereo Camera), it promised images with visual resolutions as high as 14 MPP, even better than the 19 MPP visual light camera that had travelled to Mars on NASA's *2001 Mars Odyssey* spacecraft. But even better than that, it could take images in full color as well as stereo, meaning that 3D anaglyphs could be generated from the data,

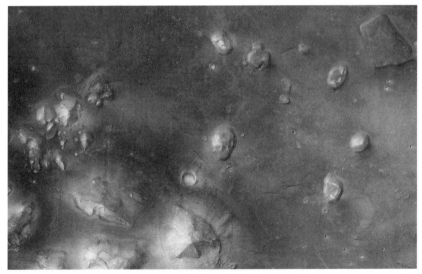

Grayscale sectional enlargement of ESA-HRSC image of Cydonia showing the City, Fort, Face, the "Symmetrical Mesa" and the D&M pyramid.

giving us our first legitimate 3D views of the Cydonia landscape.

After several unsatisfactory attempts to image the Cydonia region in the December 2004 photographic window, on September 21, 2006 the European Space Agency finally released the first HRSC color imagery of the Cydonia region of Mars. Unlike its earlier partial release of the 2004 data, this one came captioned and included a politicized press release that emphasized the usual non-arguments against the Face on Mars being artificial.

Taken on July 22, 2006 under much better lighting conditions than the 2004 images and from straight overhead, these 13.7 meter per pixel images (designated 305-230906-3253-6-col and -co2) provided the best overview yet of the area that had come to be known as the "Cydonia complex." What they reveal is a stunning landscape that is strongly confirmative of almost all of my expectations and the independent investigators' previous predictions. In many ways, because they are color, these two stereo images are far more illuminating than supposedly higher resolution images generated by *Mars Global Surveyor*. In fact, it was quite clear from examining the new images that the previous *Mars Global Surveyor* images of the Face (and Cydonia) leave a lot—quite a lot—to be desired.

Let's start with a brief review of the concept of "spatial resolution" in remote sensing (satellite) imagery. Most of us assume that an image with a stated resolution of 1.2 meters per pixel, like the HiRISE camera on *Mars Reconnaissance Orbiter* can sometimes provide, is automatically "better" than an image of 13.7 MPP, such as these *Mars Express* images. Most of the time, that's true. But there is a lot more to it than that. If the 1.2 MPP image is grayscale, meaning 8-bit data, it by definition carries less information than a 16-bit or 32-bit color image. Furthermore, all kinds of conditions—atmospheric haze, lighting (sun) angles, camera settings, the optical properties of the camera, the filters being used, the incidence angle of a nadir-pointing camera—can all dramatically affect the quality of the resulting image.

A good case in point would be the infamous "Catbox" image of the Face on Mars.

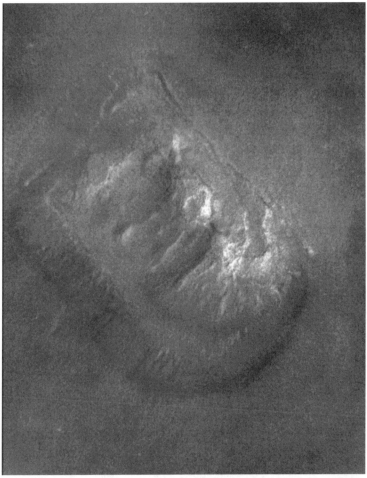

TJP enhancement of the so-called "Catbox" image of the Face on Mars.

The example above is the so-called "TJP enhancement" of the infamous (and fraudulent) "Catbox" image of the Face, taken in 1998. While it is unquestionably the best enhancement of this image to date, it is of very poor quality. According to the Malin Space Science Systems website, the image has a spatial resolution of 4.3 MPP, making it by far the best image of the Face to that point. However, this *stated* resolution only takes into account the maximum possible resolution, based on the camera optics and the altitude above the target. The image was in fact taken *after* the spacecraft had already passed over the Face, looking back up to the north, from a 45 degree angle to the west, and with the sun at a fairly low morning sun angle of 25 degrees above the horizon, lighting the Face from below. In addition, MSSS stripped out at least 50% of the data by using an exceptionally large image swath (see *Ancient Aliens on Mars*), and haze and cloud cover made for very poor lighting conditions. The result was an extremely dark, low contrast image which didn't come close to the imaging capabilities of the MGS camera under ideal conditions. Vince DiPietro, an early pioneer of Face research, concluded that with all the factors included, the effective spatial resolution of the image was 14 MPP, as opposed to the stated 4.3 or the optimum 1.2 MPP capability of the camera under ideal conditions and altitude.

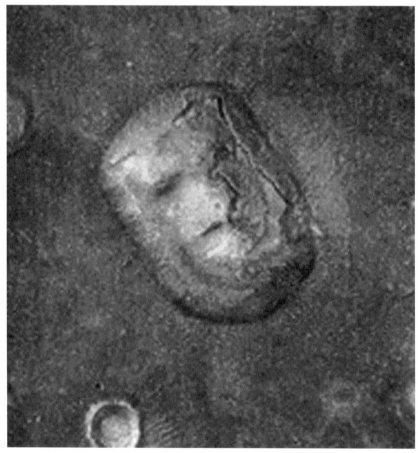

Mars Express HRSC image of the Face on Mars.

By contrast, the new ESA images were taken from directly overhead, at close to minimum altitude, under full daylight conditions with virtually no cloud cover, and in 24-bit color. Beyond that, unlike any of the previous missions, the HRSC is able to take images virtually side by side, one after the other. The results were impressive. The two stereo images, taken just moments apart, provide the best overview of the Face and City we have received from any mission so far. That still stands today. Both frames captured the Face as well as other objects of interest in Cydonia like the Fort, the D&M and the City in high resolution color.

The first image (305-230906-3253-6-co1) provides the added bonus of capturing the Cliff in the lower portion of the frame, and a substantial number of the anomalous "mesas" north of the Face first noted by Hoagland in his book *The Monuments of Mars*. What this all amounts to is that these images were without a doubt the best wide-angle view we had ever gotten of Cydonia.

So, that said, what did they tell us?

For starters, it's immediately obvious that Cydonia is a very weird place, at least as far as any natural explanations for its formations go. Most of the "named" formations that we are familiar with look just as we would expect them to. Other formations did not seem to fare as well under higher resolution, especially the Fort. The *Mars Express* color image is by far the most detailed view of this object yet, despite the fact that the grayscale MGS image has a higher theoretical spatial resolution.

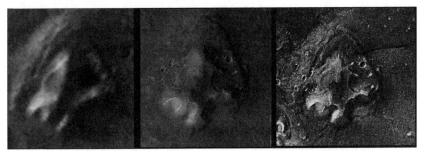

Three views of the Fort from *Viking* (left), *Mars Global Surveyor* and *Mars Express* (right).

At first blush, the newer images of the Fort and some of the other Cydonia anomalies would seem to undermine arguments for their artificiality. But, the Fort can be clearly seen to rest on a triangular platform, with a collapsed peak that must be triangular in nature, since in the *Viking* data it left a distinctly triangular shadow which gave the illusion of an "inner wall" at a 60 degree angle. Simply put, a triangular, pyramid shaped shadow cannot be produced by anything except a triangular, pyramid shaped peak.

Giza pyramids projecting long shadows at low sun angle.

An example of this would be the Great pyramids at Giza. At low sun angles, similar to the conditions at Cydonia when the *Viking* images were taken, they project long, pointed triangular shadows identical to the pointed, triangular shadow cast by the peak of the Fort. The result is not an optical illusion, but rather an accurate reflection of the shape of the tip of the pyramids. It therefore follows that the Fort has an identical pointed, artificial peak casting a similar shadow.

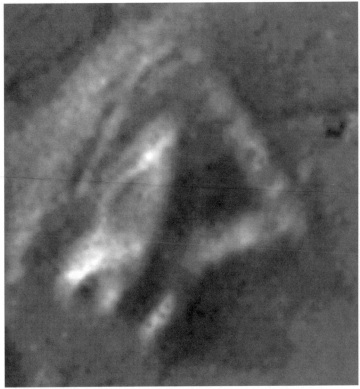

The Fort as seen in *Viking* image 35A72 (NASA).

So what then is "the Fort?" In my opinion, it is a collapsed tetrahedral pyramid. In looking at the later MGS data, it's clear that it has a triangular base which served as the foundation, and was probably supported by a central column holding up the peak and three spars at each apex. This would leave an enormous interior volume for living space, something like the Luxor hotel in Las

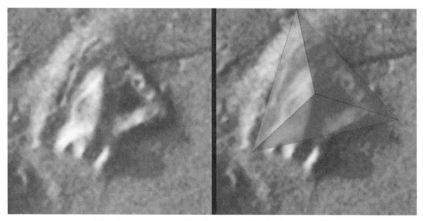

The Fort compared with a tetrahedral pyramid.

Vegas, and would fit with the model that these ruins are architectural ecologies, self-contained pyramidal cities capable of supporting thousands of inhabitants. At some point in the distant past, the

"Regular" tetrahedral form imposed over *Mars Express* image of the Fort in Cydonia.

106

structure was inundated by a massive landslide from the northwest, which broke down and bent over the pillars, allowing the central peak to collapse straight down but maintain its basic pointed shape.

This impression is reinforced when a tetrahedral form is superimposed over an image of the Fort itself, showing that the three straight lines which make up the equilateral "base" of the tetrahedron match almost perfectly with foundation of the Fort. It's also an observation which is only strengthened by the higher resolution and more directly overhead views produced by *Mars Express*. I cannot emphasize how unlikely it is that these proportions can be found in a naturally occurring landform. Erosion simply does not produce naturally perfect equilateral triangles except under the rarest and most bizarre of circumstances, and *never* within a stone's throw of massive pentagonal pyramids and 1.5-mile wide "naturally" sculpted human faces.

The D&M Pyramid, just a few miles south of the Face and the Fort, is yet another example of how the *Mars Express* HRSC camera has added to the mystery and archaeological intrigue of Cydonia. Under its high resolution gaze, the D&M's artificial construction becomes fact.

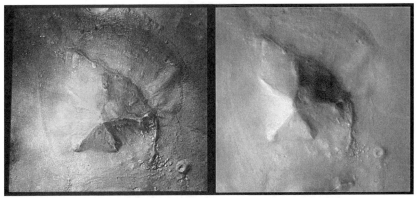

Two views of the D&M pyramid from ESA's *Mars Express* (left) and NASA's *2001 Mars Odyssey* (right).

The *Mars Express* view shows the object in much better context, emphasizing the strange circular platform from which the pyramid rises some 2.500 feet above the plain below. The *Mars*

107

Express image is also dramatically better than the *2001 Mars Odyssey* visual camera image, showing more detail than can be discerned in that grayscale image while losing none of the critical context. In this detail, we can see the lines of bilateral symmetry, the "exit wound" to the east and the inward collapse of at least one facet of the pyramid. But it is in the 3D stereo images that the HRSC camera really pays its way.

The 3D views are an intended benefit of the stereo capabilities of the HRSC. Created pretty much automatically from the existence of two nearly identical datasets (images co1 and co2) these perspective views offer us the ability to "fly over" Cydonia at a resolution never imagined before. What they show is the overtly— but fractally eroded— pyramidal form of the D&M. Arguing that this formation is anything but a construct becomes absurd in the face of data like this.

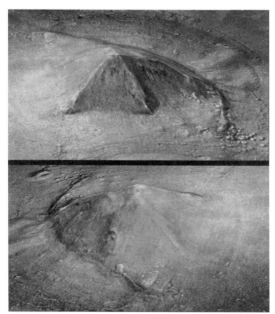

3D perspective views of the D&M pyramid from *Mars Express.*

But perhaps no member of the existing "Cydonia complex" is revealed as more alien than the curiously straight edged "Cliff."

The "Cliff."

First noted by Hoagland back in the early days of the Cydonia investigation, the Cliff is an anomalously straight ridge seemingly perched atop the ejecta blanket of a nearby crater. It formed, along with the tetrahedral apex of the "Tholus" and the rim pyramid of

109

the nearby crater, one of the key 19.5 alignments in the Cydonia Geometric Relationship Model (see *Ancient Aliens on Mars*). The argument for its possible artificiality has always centered around the fact that it has not only a completely bizarre geomorphology (a near-perfectly straight ridge running for miles) but that it rested on a raised platform of sorts which seemed to postdate the crater impact. Later high resolution images from *Mars Global Surveyor*, *2001 Mars Odyssey* and now *Mars Express* seem to support this. There is no evidence that the ejecta has spattered on top of the Cliff, and the entire mesa just seems to have been "stamped" on top of the ejecta blanket. After more than a generation, the conclusion that this anomalously straight ridge somehow postdated the cratering event (which is a geological impossibility) would have to be considered validated.

Nearly symmetrical mesa with nearby ruined tetrahedral pyramid just southwest of the Face on Mars.

The new images also cast light on another "object of interest" in Cydonia just southwest of the Face. Described in the ESA article accompanying the new Cydonia images as a "skull shaped mesa," this object has fascinated investigators since it was first noted on the original "Catbox" image strip.[1] The new ESA image showed that what appeared in the *Viking* data to be simply a natural mesa among the ruins of Cydonia was in fact an intriguingly symmetrical object more than a mile across.

The odd ESA description of it as "skull shaped" is attributed to the claim (by ESA) that some people have referred to it as such. In fact, I have never encountered this description in any anomaly related web article or public posting other than the ESA press release. The "mesa" in question has caught my fancy for several reasons: its significant degree of symmetry, its proximity to a tetrahedral ruin

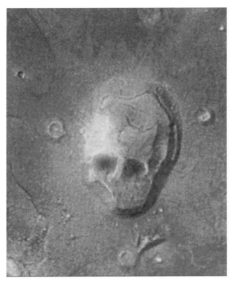

"Skull shaped mesa" with tetrahedral ruin to the south.

about the size of the Great Pyramid of Giza noted by Hoagland in 1998, and the existence of a tunnel or channel terminating at the exact lateral center of the "mesa" in the *Mars Odyssey* infrared data. My own conclusion is that the skull reference is a designed distraction, to keep readers from viewing it from a perspective that makes the symmetry more obvious.

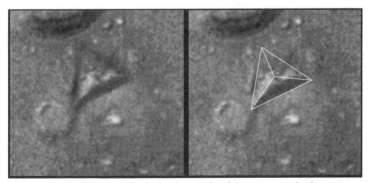

Close-up of the tetrahedral ruin south of the symmetrical mesa.

This symmetrical mesa, the Fort and the D&M, along with a series of anomalous objects north of the Face and City are excellent examples of a phenomenon called "Fractal Erosion." In mathematics a fractal is a dataset that displays "self-similar" characteristics.

In other words, it assumes that a given object resembles itself at different levels of magnification. For instance, the Face on Mars still looks like a Face at both lower resolution (*Viking*) and higher resolution (MRO). With fractal erosion, the concept is taken a step further to predict that a given object also generally *erodes* along the lines of its original shape. So we can reverse the erosional process and reconstruct the original shape of a given object by studying how it appears today. Using this methodology, satellite imaging expert Erol Torun did a study of the original shape of the D&M pyramid in the early 1990s. We can do a similar reconstruction with various objects around Cydonia. Doing so, the "skull shaped mesa" becomes more overtly symmetrical, and the original shape of a number of objects in the north become highly geometric.

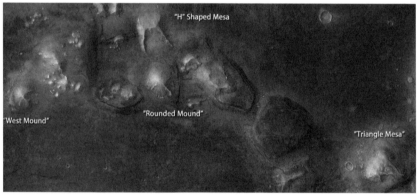

"Super Bowl City" north of the Face on Mars in Cydonia.

This area, dubbed "Super Bowl City" by researcher Robert Harrison of the Cydonia Quest web site, was first noted by Hoagland in *Monuments*. It contains a wealth of geometric ruins that are nearly inexplicable by natural erosive processes. In fact, all of the "mesas" in this particular area are very unusual, but the four I have selected seem to be the most non-natural and the most consistent with the other anomalies at Cydonia.

The far left object, dubbed the "West Mound" by Harrison, is markedly similar to other objects at Cydonia like the D&M and Main Pyramid in the City. It appears to have a degree of symmetry about a central axis that is highly unusual for any natural erosion process. Close up MGS views reveal blocky, room sized geometric patterns on its more eroded flanks, where it appears the original exterior structure has collapsed inward. If this object was considered part of the original "City" it would certainly have been included as a candidate for artificiality, and would certainly have passed the test upon closer inspection.

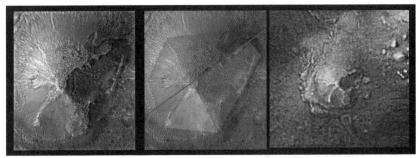

Three views of the West Mound.

Just a short distance from the West Mound are two more objects that really defy conventional explanation. One is a bizarre "H" shaped formation.

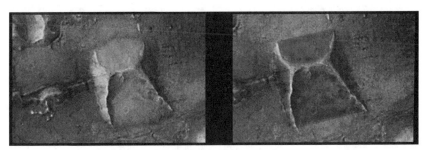

The "H" shaped formation.

Once again we see the unaccountable symmetry of a "mesa" in Cydonia, this time in the shape of an "H." The fractal reconstruction shows it was once a wedge shaped structure, probably a hollow arcology which has collapsed inward around an "H" shaped frame. The superstructure is clearly sagging and caving inward, but just as obviously it was once a far more robust and majestic building. Directly adjacent is what looks to be a bulging "bunker" with a vertical face and what could be judged as entrances. Assuming the reconstruction is valid (and it would seem to be obvious), then these two adjacent structures would easily be deemed artificial. If they appeared on images anywhere on Earth, it is doubtful that their artificiality would be at all questionable.

Just south of this complex is another object nicknamed the "Superbowl" by Bob Harrison. Rather than a bowl, the object in question is actually a rounded mound about the size of some of the pyramids in the City. In the ESA images, a distinct "moat" can be seen around the circumference of the mound, indicating it has sunk at least partially into the surrounding terrain.

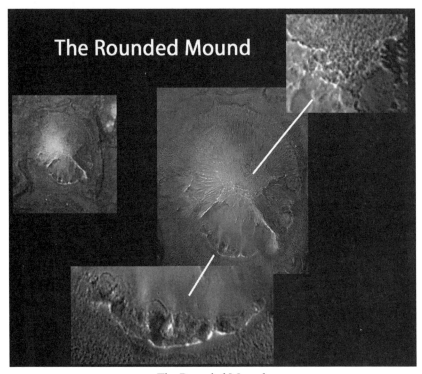

The Rounded Mound.

Close-ups done by Harrison show that there are large pits or sinkholes around the base of the structure, implying that it is settling into a hollow or collapsing inward from the base into a hollowed out interior. Obviously, neither of the characteristics is consistent with a naturally eroding object, unless it is some bizarre example of volcanism which forms with air pockets inside. This notion is pretty quickly dispelled by looking at the exposed upper surfaces of the mound, which display the all too familiar rectangular "room sized cells" where the casing structure appears to have been worn away.

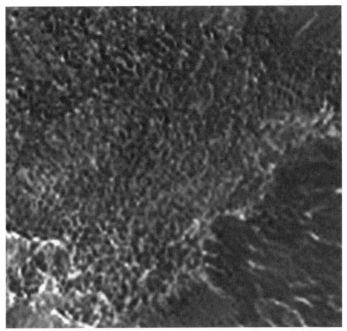

Close-up of exposed "room sized cells" on top of the rounded mound formation.

This is consistent with observations from the D&M and the so-called "main pyramid" in the City. All of them display these regular, repeating geometrically arranged cells. They are confirmation of the core idea that these are constructed objects.

The best, however, is saved for last. Just a little further east of this collection of objects is a very strange triangular shaped "mesa" which seems to have three matching geometric nodes at each of the triangle's three vortices. It is eroded, to be sure, but what kind of natural process erodes a simple hill into a base equilateral triangle with such overtly geometric nodes at each corner? This "mesa" is as anomalous as it gets, and for Cydonia, that's saying something.

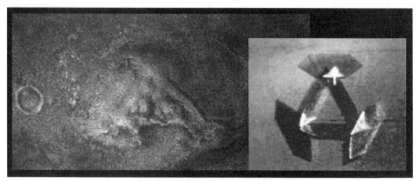

The Triangle Mesa comparison.

As we get into the details of the Triangle Mesa, there is even more strangeness. Just to the left of the northernmost node is a very odd dark feature which appears in both the December 2005 and July 2006 datasets—a distinct "T." This very dark marking on the structure is not only aligned perfectly north\south—as is the triangle mesa itself—but its two dark lines intersect at precisely 90 degrees. A pretty cool trick for a "naturally eroding" mesa.

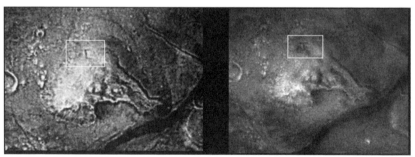

Perfect "T" shaped formation in two different images of the Triangle Mesa.

The more recent image suggests that these dark, interconnected lines may demark the edges of the eroding upper node, perhaps where the base of the node has slumped inward, albeit in a very precise and geometric fashion. With its strange markings, triangular base and geometric nodes, this Triangular Arcology is more than anomalous enough to be added to the candidates list for artificiality at Cydonia.

This brings us, at last, to the Face itself.

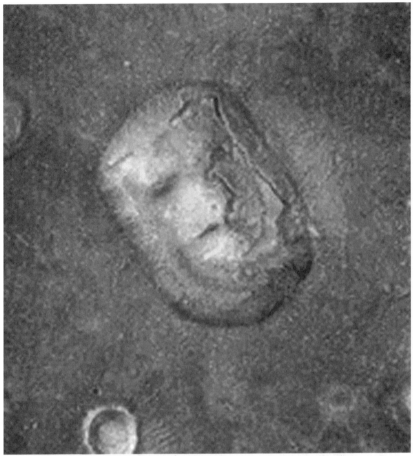

Averaged composite image of the Face on Mars from ESA images 305-230906-3253-6-co1 and 305-230906-3253-6-co2.

You would think, after all the images now taken of this enigmatic object, that there would be little if anything new we could learn about the Face. However, despite the fact that there are now some 25 partial or complete views of it from five different missions and cameras, the mystery of the Face endures. Not only do these new HRSC color images tell us new details about the Face itself, they serve to point out the problems and issues with earlier images of the Face, even at the supposedly "higher resolutions" we discussed earlier.

The first thing we can note is that while the East and West sides of the Face (or "City and "Cliff" sides, as they have come to be known) are made of two distinct materials (see Keith Laney's Cydonia IR images), their surface color is predominantly uniform. This indicates that whatever is covering most of the Cydonia region is most likely a layer of reddish dust. This also completely refutes Dr. Philip Christensen's assertion in July of 2003 that the Cliff side was covered with a dense snow pack, thus accounting for the dramatically anomalous reflection seen in the pre-dawn THEMIS color image from 2003 (see *Ancient Aliens on Mars* and chapter 4 of this volume). If this dense layer of snow existed, we'd see it in the color image as a bright white casing around the base of the Face. Since this does not appear, we can safely judge that the anomalous brightening in the THEMIS image was due to another (most likely artificial) cause, as I have already outlined.

Another thing these images provide is confirmation of earlier observations. Secondary facial characteristics—which flatly cannot exist if the arguments for a natural origin of the Face are valid—are reconfirmed in the ESA images. We can plainly see the two "nostrils" in the nose, which first appeared on the infamous "Catbox" image in 1998, then seemed to disappear in the low contrast MGS images taken from above the "forehead" of the Face in 2001 and later.

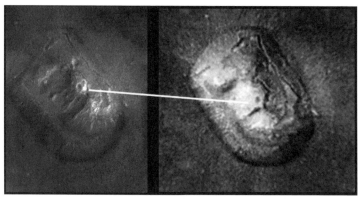

The "Nostrils" in the nose of the Face on Mars from MGS (left) and *Mars Express* (right).

Their reappearance is due to a simple function of spacecraft geometry. These are the first images taken from almost directly overhead, as opposed to the many MGS images which were substantially up-track or down-track from the Face's actual latitude.

In addition, we can see the sharply defined brow ridges on both sides of the Face as well as the overall symmetry of the base platform. Indeed, the most compelling observation that *Mars Express* provides is that the two eye sockets precisely align straight across the Face. This is in sharp contrast to the MGS images over the last couple of years, which have shown the eye sockets to be substantially out of alignment and the Face platform to be substantially wider than it actually is.

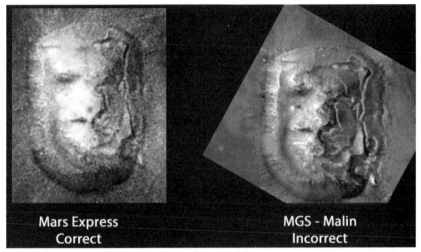

Mars Express
Correct

MGS - Malin
Incorrect

Comparison of orthorectification of *Mars Global Surveyor* image of the Face on Mars with *Mars Express* image (left).

This dramatic difference is due to two factors. First, as mentioned above, is the more directly overhead image angle taken by the *Mars Express* instrument. Second, the improper orthographic rectification of the *Mars Global Surveyor* images contributed substantially to this distortion and the overall illusion of asymmetry. In fact, the Malin orthorectification is so bad that it widens the base, twists the nose and pulls the right side eye socket significantly below that on the left side.

We can see with a side by side comparison just how far off the Malin version is. However, if the Malin/NASA version is bad, then the orthorectification produced by SPSR's Dr. Marc Carlotto from the 2001 MGS image is positively abysmal. It is even more stretched, distorted and un-face-like than Malin's, and it clouds the situation even further than the poor job NASA did.

Orthographic rectification of MGS Face on Mars image by Dr. Mark Carlotto.

No wonder nobody at NASA seems to "see" the Face the way we do.

What all this illustrates is that orthographic rectification is something of a black art, and even those who would claim to have it mastered can be exposed in the light (and color) of the day. It would seem the questions of the position of the two eye sockets and the existence of the "nostrils" are forever settled, and once again the independent researchers have won the day. None of this, however, seemed to move the powers that be within the mainstream astrophysics community. Case in point: the ESA.

The European Space Agency's release of the Cydonia data was accompanied with a typically shallow and half-hearted "naysaying" article, claiming once again that there was nothing at all unusual about the Face, and that gee, it wasn't a Face after all. I've gotten used to these silly political documents accompanying new image releases from NASA, although they usually resort to far more ruthless and dishonest propaganda techniques than did the ESA (see *Ancient*

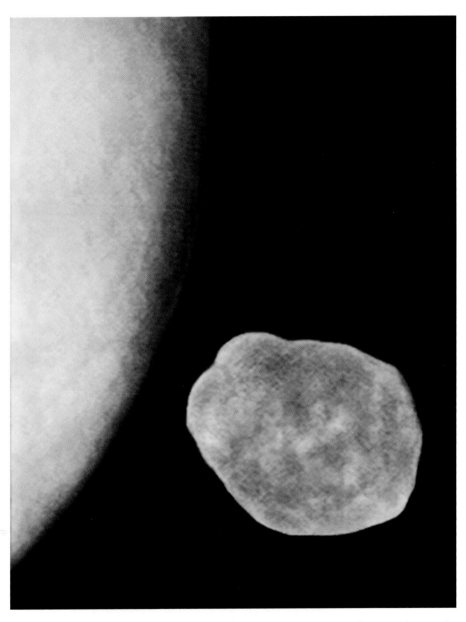

Phobos II image of Mars' moon Phobos showing anomalous right angle mesh pattern all over the body of the moon.

2 color views of the THEMIS infrared images
of Cydonia. The "Fort" is in the upper left.

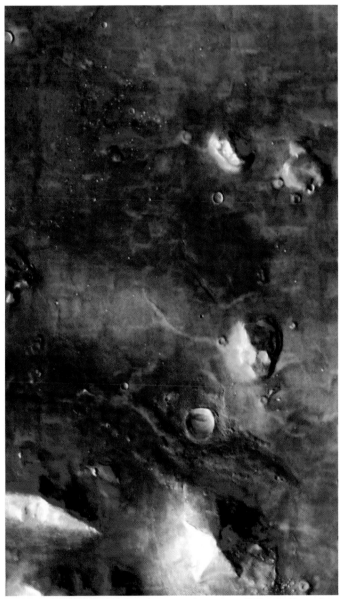

Processed THEMIS color image showing "Blocks" under the surface of Cydonia. Blocks follow true Cydonia North/South, not the image scan North/South.

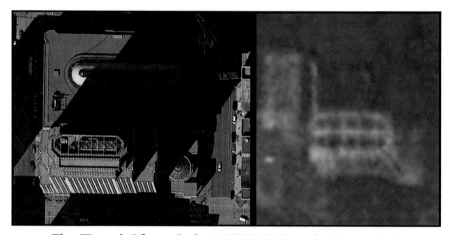

The "Temple" from Cydonia THEMIS IR and a Terrestrial comparison (Minneapolis)

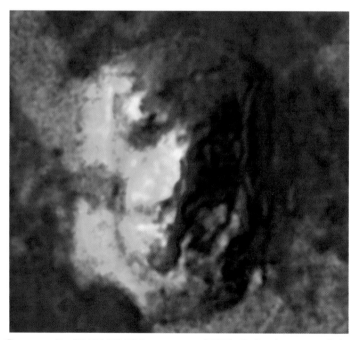

Composite THEMIS IR image and ESA Color image of the Face on Mars. Note structural detail on right side of the Face, eyeball detail & implication of teeth in the mouth.

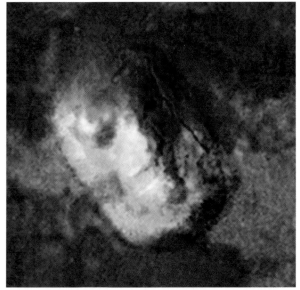

THEMIS IR/Mars Odyssey 2001 visual light
image composite.

Regular tetrahedral form imposed over ESA color
image of the Fort.

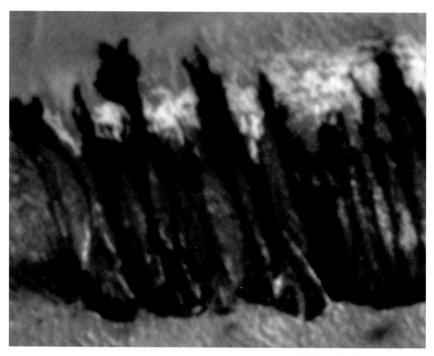

Swaying "Trees" and a moss covered rock on Mars

Mechanical parts from all over Mars

The "Cylinder" and the "Tank" from PIA 16700

Aliens on Mars). The ESA article was pretty mild by comparison, and was obviously intended for Face novices and the Space.com crowd, who can be counted on to never ask a hard question of their authority figures. Regardless, the fact remains that after 30 years of back and forth debate and discussion of the entirety of the Cydonia artificiality hypothesis between the independent researchers and the NASA\ESA establishment, the mainstream argument still comes down to the same thing it did in 1976: "It's not a Face."

Or, more accurately: "It's not a Face, in spite of the fact it rests on a bilaterally symmetrical platform, it has two aligned eye sockets, the tip of the nose is the tallest point on the structure, there are two clearly defined nostrils in the nose, the west eye socket is shaped like a human eye including a tear duct, there is a spherical pupil in the eye, there are rectangular, cell-like structures around the eye, the two halves of the Face make up two distinct visages when mirrored (one human, one feline), it is placed nearby a series of pyramidal mountains which have rectilinear cells visible in the interiors at high resolution, it is in close proximity to a pentagonal 'mountain' which is bilaterally symmetrical about two different axes, it has anomalous reflective properties under pre-dawn conditions, it sits atop a sheet of ice covering a vast network of underground lines and blocks that closely resemble a large city, it is surrounded by a series of tetrahedral mounds which are placed according to tetrahedral geometry, it is within shouting distance of a series of newly observed objects which include a triangular 'mesa' with geometric nodes at each vertex, it...."

Well, I could literally go on and on. But you get the point. Their argument is really weak.

Perhaps that's why, with each of these releases, they take such care to talk about the Face as if it is an isolated anomaly. It's weird enough by itself, as we have seen. But when you start to add all the other objects into the equation—the City, the Fort, the Cliff, the Tholus, the D&M—the mainstream argument, weak as it is, completely collapses.

Maybe that's why this time, ESA, or at least HRSC principal investigator Gerhard Neukum, wasn't quite up to the disinformational task. Maybe that's why they needed a little help from their friends at NASA.

Even though they were very high quality, both of the new HRSC images were displayed in a rather bizarre manner. In almost every other case, the convention for such image releases is to display the images with north being the "up" direction, south "down," and west and rast being left and right respectively. Instead, the ESA images are displayed with north to the right, effectively altering the scientific convention and forcing anyone seeking to study the images to rotate them 90 degrees counterclockwise to see the Face and Cydonia in their normal orientation. This had the effect of disorienting casual readers, who either had to rotate the images in an image editor or turn their heads to "see" the Face right side up. Again, the only logical purpose of such an annoying change in convention is to suppress the interest of the casual reader, since the spacecraft was in an 86° inclined polar orbit, meaning that the data would have come back to Earth with north "up" in the first place. The ESA imaging specialists actually had to rotate the images 90° to the right to make it look the way it was displayed.

Given that, one has to question the integrity of the people putting the article together. If the mainstream argument is so strong, if the Face and other objects at Cydonia are products of "simple erosion" as Agustin Chicarro, ESA's chief scientist for *Mars Express* argues, then why resort to the confusion tactics? Why force the reader to download and rotate the image just to look at it from the same perspective it has traditionally been seen in? Really, if the process was honest, there is no reason to do so.

But the inherent weakness of their arguments, and the lengths they will go to in order to preserve them, could not be more dramatically represented than in the stunning 3D perspective views generated from the new data. As we saw with the perspective views of the D&M, these stereo 3D projections are extremely revealing.

These new views have provided an invaluable look at several controversial features. The "massive tetrahedral ruin" for instance, generally dismissed by the anomalist community at large, is

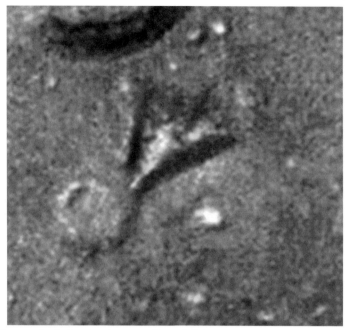

The tetrahedral ruin close up. Note bright structural "girders" on left side, just forward of remaining casing wall.

shown in fine detail for really the first time. The two perspective views allow us to look inside the object and clearly discern that Hoagland's original thesis about the object—that it used to be a full tetrahedral pyramid—is not only a valid speculation but highly probable reality. The one surviving face of the former tetrahedron is indisputably triangular, or once was.

This is quite a neat trick for a product of "simple erosion." Not only that, but the facing appears to overhang the interior support structures, as if the casing was once exactly that. There are several clearly visible structural members (girders?) in the interior of the ruin. It's fairly easy to follow the lines and the remaining partially buried walls and reconstruct the original shape from them.

Truthfully, there are so many strange objects on the ground at Cydonia that it would take years for us to fully go through and analyze them each to the level they deserve. They simply cannot be explained as natural geology.

But for each new insight and discovery that this 3D process can provide, there is a countermanding dark side to it. Just as we

The "Apollo Patch" Crater

An often overlooked (but very weird) "crater" in Cydonia.

have seen NASA do before them, the ESA has used (or rather, misused) the 3D process to misdirect its readers. Once again, the fraud has to do with the Face itself.

The authors of the ESA article obviously realize that while there are dozens of anomalous objects at Cydonia, they have no hope of discounting all of them to readers with any common sense. As a result, they resort to the time honored canard of reducing the entire Cydonia artificiality model to a single question—that of the Face on Mars.

They realize that as the first anomaly to be noted at Cydonia, the Face is the cornerstone from which the rest of hypothesis has sprung. Their thinking is that if they attack the Face relentlessly and without regard for the truth, it won't matter how much evidence there is supporting the other objects in question. Apparently, it is this reductionist strategy which drives them to fabricate data as they did near the end of the article with a 3D perspective view of the Face.

Somehow in this new 3D perspective view the Face on Mars has managed to acquire a distinct lump between the eyes. Like some interplanetary Elephant Man, the Face has miraculously morphed into a grossly distorted version of itself. I knew immediately that the lump was a fraud, just like the Catbox, MOLA or Middle Butte Mesa NASA scams (see *Ancient Aliens on Mars*), because the Face had been imaged numerous times at sun angles and resolutions that

124

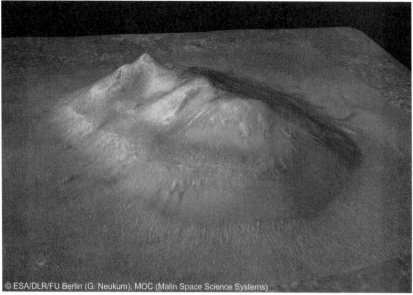

© ESA/DLR/FU Berlin (G. Neukum), MOC (Malin Space Science Systems)

The "Elephant Man" version of the Face on Mars.

would have revealed any such lump if it was a real feature. They had not.

At first, I was perplexed by this, because to this point in the political article ESA and Dr. Neukum had at least shown us what appeared to be honest data. But now, with this newly acquired protuberance, Neukum and ESA had gone from a fairly neutral stance to jump into the camp of the worst kind of scientific charlatanism (think Phil Plait or James Oberg). What had suddenly happened to convince Dr. Neukum to stoop to such an obvious fabrication?

Then I saw the 3D image credits: "ESA\DLR\FU Berlin (G. Neukum), MOC (*Malin Space Science Systems*)." (Emphasis added)

So I had my answer. Once again, "Malin happened."

The images with the distorted hump had been generated with "help" from the boys at MSSS, the heart of darkness when it comes to Cydonia. So it then all made sense. Using Malin's (probably deliberately) bad data as a constraint, the ESA had generated the Elephant Man version of the Face on Mars.

To his credit, on his *own* website, Dr. Neukum had not used the MSSS generated version but instead generated his own 3D perspective views of the Face purely from the HRSC data. Not too surprisingly, they did not include the now infamous "hump." But what is even more intriguing is what they *did* include…

3D perspective view of the Face from European Space Agency HRSC data. Note the complete lack of the "hump" from the version produced with NASA/Malin data.

In Nuekum's 3D images of the Face, certain details can be seen where none had been seen before. In all previous close-ups of the Face taken by Malin Space Science Systems, the area around the lower left (western) "chin" and the lower left section of the base slope are mysteriously absent of any significant detail (see the "hump" image, above).

This results in odd blurry areas on this part of the Face, similar to other strange blurry areas I have noted on some other images of Cydonia over the years. Histogram comparisons show that these areas contain far less detail than other portions of the images. Curiously, these blank-blurry areas seem to be quite rare in the MOC images of other parts of Mars, but appear to be quite common in the Cydonia images. In short, I've never trusted them.

What Neukum's new color images—both 2D and 3D—now

126

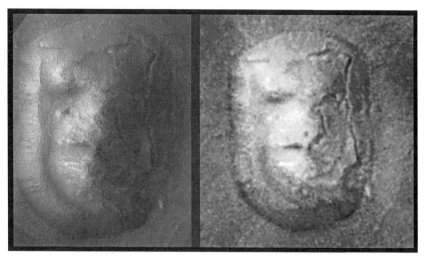

The Face from *Mars Express* (scaled up, right), and from MRO (scaled down, left). Note that the blurry areas in the supposedly higher resolution MRO image are less detailed than the comparable areas of the *Mars Express* image.

convince me of is that I have been right to be suspicious all along.

A side by side comparison of Neukum's 2D color image and the latest "high resolution view" of the Face from JPL's *Mars Reconnaissance Orbiter* show the that there are significant discrepancies in the details of these areas of the Face. In the MRO image, as with all of the NASA\JPL\MSSS images of the Face, we see these odd, cloud-like blurry areas around the southwestern portions of the Face. In the *Mars Express* images (which are the only images of the Face from cameras not controlled by NASA) these areas appear much like the rest of the Face, showing fine structure and hinting at the possibility of structural details which might be seen at higher resolutions. In fact, the 3D image of this part of the Face from Nuekum's own website shows these areas specifically, and the "cloudy bits" are nowhere to be seen. In fact, what you do see are structural rebar, tubes, girders and the like, reaching up from the Cydonia plain to attach themselves to the base platform of the Face.

The area around the chin is less distinct, but you can plainly see it is not the drab, featureless blur always depicted in the NASA data. Now, I would not begin to compare the theoretical spatial

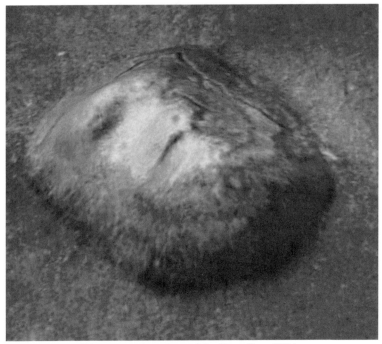

3D perspective view of the Face from ESA.

resolution of the MRO camera to that of HRSC on the *Mars Express*. But isn't it interesting that the only dataset which is not under the control of Michael Malin and his NASA/JPL buddies is so discrepant from the NASA dataset in this area? And considering that the western half of the Face is bound to be more eroded (due to the predominantly western winds of the region), wouldn't this side logically be the area where the underpinnings of the Face's internal artificial structure, if it had them, would be most obviously exposed?

You bet it would.

So, is that why they had to produce the "Elephant Man" version of the Face, and blank out the other areas in the MRO data? To keep anyone from noticing what's really down there? Will we be forever dependent on outside sources for "real" data on Cydonia, or will NASA and JPL ever give us the clean, straight up ground truth on Cydonia? And how subject to pressure will NASA's partners like the ESA (who felt compelled to accompany this new image release with the usual political tripe about the Face) be?

It's sad to have to paraphrase former Secretary of State George

Shultz again, but the "Elephant Man" incident doesn't change my opinion of NASA's integrity; it only confirms it.

So much time and space in my two Mars books has been dedicated to the Face and Cydonia, we have almost lost track of how many other areas of Mars have mystery and intrigue to offer. Finally, it's time to look at those and the work being done by a veritable army of independent researchers who have also found that Mars is significantly more interesting than NASA would have us believe.

(Endnotes)

1 http://www.esa.int/Our_Activities/Space_Science/Mars_Express/Cydonia_-_the_face_on_Mars

Chapter 6
The Martian Zoo

One of the alternative ideas presented in the study of Martian anomalies has been the existence, or at least the alleged existence, of large scale "glyphs" on the surface of Mars. Over the years,

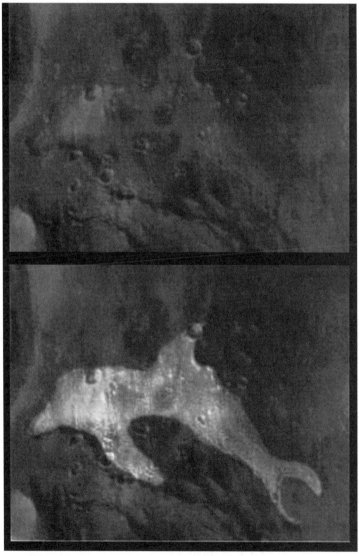

The Cydonia "Dolphin."

many researchers ranging from the serious to the silly have proposed that certain formations were meant to represent various terrestrial figures and animals. Among these are a dolphin (which is actually Hoagland's "massive tetrahedral ruin" near the base of the symmetrical [skull shaped] mesa just southeast of the Face), a seahorse, a scorpion, several different types of birds and a number of human looking figures. Researcher George A. Haas of the Cydonia Institute has specialized in these studies, compiling them into several books about the formations around Cydonia and Mars.

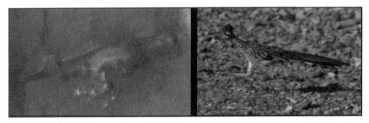

The "Road Runner Complex" in north Cydonia (George A. Haas).

This actually all started shortly after the first *Mars Global Surveyor* images of Cydonia and the Face came back in 1998. At that time, several of the self-described "serious" Cydonia researchers announced that they had spotted the "Dolphin" in the image strip containing the Face on Mars. These "serious" Cydonia researchers (as opposed to me and Hoagland, who were branded as "conspiracy theorists" because we dared to point out NASA's ongoing mendacities regarding everything about Mars) also saw— or thought they saw—numerous other strange images and anomalies in the new images of Mars. The "pictograph" craze really took off from that point.

That particular image also contained something else that these "serious" Cydonia researchers were quite excited about; what they decided was writing, and by that I mean Hebrew-Arabic script, on one face of the D&M pyramid. Both Hoagland and SPSR's Dr. Tom Van Flandern had spotted the symbols at virtually the same time. Hoagland wanted to go public with the information, while Van Flandern wanted to hold the writing on the D&M for future use. In the course of the discussions, Hoagland discovered that the

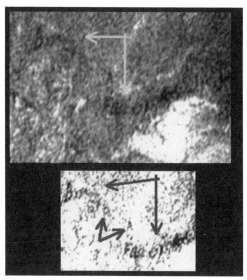

"Letters" on the D&M pyramid.

official position inside SPSR was that these symbols were features embedded on the D&M Pyramid itself. They steadfastly refused to consider the conspiratorial alternative—that someone had *placed* these symbols on the image. As a result, they were forced into a totally absurd scientific position.

Let's consider this for a minute. If there are really letters on the D&M, then that has all sorts of implications about who put them there. The first assumption must be that Martians— hundreds of thousands if not millions of years ago, at least—used Hebrew-Arabic lettering in their ancient communications. And not only that, but they scrawled these modern English language letters on the D&M, like some sort of cosmic graffiti, just for us to find. Further, these "symbols" just happened to be oriented on the pyramid in such a way that they could be read without even having to rotate the image to make them appear right side up.

A far more plausible explanation is that somebody at Malin Space Science Systems or JPL simply put the letters there. These letters, which are on the original image on the official NASA site, were put on the D&M, an object of obvious interest to us, as a message. They are clearly meant to confirm that the image had been altered.

Whether this was done as a whistle blowing move, or simply to rub in the impunity with which MSSS and JPL felt they could manipulate the data, this is clearly a far more logical explanation than SPSR's anonymous Martian graffiti artists. Yet, SPSR (through Van Flandern) refused to budge from their ridiculous stance that these symbols were genuinely on the D&M. Beyond that, since they had found them first, they requested that Hoagland not make mention of them. Out of respect for their priority, he agreed.

Of course, these letters have never appeared on any other images of the D&M. After looking at them many times over the years, I have concluded that they might say either "Barsoom" (Edgar Rice-Burroughs' name for Mars in his novels), or "Hoagland is a fag," which I am reasonably certain is the last discernable word.

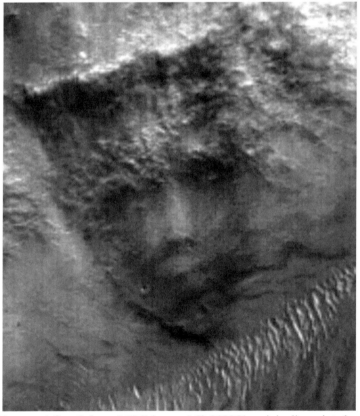

The "crowned face," from MOC image M0-203051, originally pointed out by researcher Greg Orme.

At any rate, the point is that when all of this talk of writing on Mars and images of dolphins began, I was decidedly on the opposite side of the notion. That is still my position, but I do think there is enough data of interest to examine the question more closely. SPSR and the late Dr. Van Flandern apparently felt the same way.

On April 5, 2001 Van Flandern held a press conference at the National Press Club in New York and pushed the "pictogram" debate into the mainstream discussion of Mars and Cydonia. Not only did he present the "dolphin" and the writing, he also added several more that hadn't been much discussed previously, including the so-called "crowned head face."

The image, a crop from MOC image M0-203051, was first pointed out by Greg Orme, who found it while sifting through a boatload of new data from the *Mars Global Surveyor* camera. Located on a cliff edge in a region known as Libya Montes, the so-called crowned face has been compared to sculptures along the lines of Mount Rushmore in South Dakota. However, there are several differences worth noting. First, there are four distinct faces at Mount Rushmore, whereas the crowned face is isolated, although some, including Orme, have argued that they can see the faint remnants of other faces carved into the cliffs. Frankly, I don't see those at all.

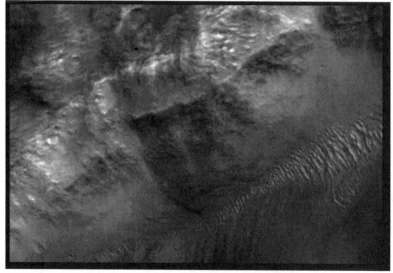

Enhanced context view of the "crowned face."

135

Van Flandern, at his press briefing, compared it to the Cydonia Face:

While not near the Cydonia area, this face portrayal is again striking for the richness of its detail, far better than the typical face arising in clouds or geological formations on Earth. The latter tend to be distorted and grotesque when they are more than simply impressionistic.

While I basically agree with Van Flandern to a point—this does look a lot more like a face than a typical cloud formation, for instance—to me these arguments really don't hold up. The Cydonia face is on the ground, looking up, and has a myriad of structural detail which this rather random collection of shapes does not. The Cydonia Face is also surrounded in context by numerous other artificial objects of all kinds, including pyramids and what appear to be man-made (or Martian-made) mounds and other structures.

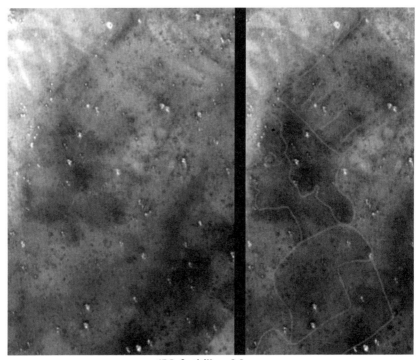

"Nefertiti" on Mars.

136

This crowned face does not, nor does it really look much like Mount Rushmore. There is little, if any, 3D relief when compared to Mount Rushmore. To me, the crowned face doesn't look any different from the rest of the cliffside it is embedded in. Rather, the viewers have simply selected and emphasized some features over others to come up with a "face." The Cydonia Face is by contrast completely different from any of the random hills or constructed pyramids around it.

Another feature that was raised in the press conference and championed by the defenders of the "pictogram" concept was the so-called "Nefertiti" figure found in the Phoenicis Lacus region. Discovered about halfway down NASA/JPL/MSSS image M0-305549 by researcher J. P. Levasseur, the figure does bear a striking resemblance to sculptures of the Egyptian princess Nefertiti, but only in profile. This raises the inevitable "Old Man in the Mountain" argument, which has been used inappropriately for years to dismiss the Cydonia Face as an optical illusion. In looking again at the greater context, there simply is nothing else anywhere near the Nefertiti figure that looks even remotely artificial. That in and of itself does not discredit it, as there are plenty of examples of terrestrial pictographic art that are isolated and surrounded by a natural environment. One example is the Paracas candelabra in Peru.

Visible from the air and as far 12 miles out to sea, the 595 foot formation is often attributed to the Paracas culture which dates from 200 BC, but its exact age is unknown. Local tradition holds that it represents the lightning rod or staff of the Inca god Viracocha (also known as Kon-Tiki) who was worshipped throughout South America and gave life to mankind. According to the legends, Viracocha "made the sun, moon, and the stars. He made mankind by breathing into stones, but his first creations were brainless giants that displeased him. So he destroyed them with a flood and made a new, better creature (Man) from smaller stones."

The Nefertiti figure in M0-305549 is much larger than the Paracas candelabra, measuring some 460 feet by 2,550 feet long. Now, in the image strip, the Nefertiti figure is oriented north/south, but without substantial image enhancement, it just doesn't

The "Paracas candelabra" in Peru.

really pop out at the viewer. By contrast, both the Cydonia Face and Paracas candelabra jump out immediately as artificial against a natural backdrop. Truthfully, the Nefertiti pictogram—if that is truly what it is—is likely to be a lot older than the Paracas candelabra and harder to make out due to millennia of erosion. But still, in my opinion you really have to strain to even see the outline of the figure, especially the "body." In short, to get to "Nefertiti," as with the crowned face figure, you have to jump through a number of empirical hoops that I just can't justify. In looking at the raw image strip, the Nefertiti figure is just a faint series of lines that don't really stand out from the background.

This brings up, once again, the touchy subject of "pareidolia."

As I wrote in *Ancient Aliens on Mars*:

In recent years, as better and better images of the Face on Mars and other anomalies on the Red Planet have become increasingly recognized as artificial, NASA and NASA backed debunkers have retrenched and attempted to hide behind a non-existent limitation of human perception they call "pareidolia." According to the debunker crowd, pareidolia is a supposed human tendency to recognize facial patterns where none actually exist. This mythical, made-up tendency has no basis in fact, has never been written up or published in any scientific or medical journal, and has failed to meet even the most basic standards of a true medical or psychological disorder.

The reason for this is simply that in the debate about the Face and other anomalies at Cydonia, they haven't a leg to stand on. The evidence that the Face and other pyramidal objects at Cydonia are artificial is overwhelming. Given this, in recent years it has now become de rigueur to redefine "pareidolia" to be a perceptual disorder whereby humans supposedly simply see "patterns" where none exist. This is akin to the man-made Global Warming alarmist crowd no longer using that term but instead citing the generic "climate change" as the basis for their socialist environmental proposals. After 17 years of cooling and several studies that link global temperatures to solar activity, the entire premise has been falsified. The same applies to "pareidolia."

In fact, the depth of the lie that is "pareidolia" can easily be found by simply tracing the word's origins. It is nothing but a phony, pseudo-scientific term first coined in 1994 by a UFO debunker named Steven Goldstein in the June 22, 1994 edition of Skeptical Inquirer magazine. This alone should tell you all you need to know about its credibility in the realm of ideas. Despite a complete lack of any valid scientific or medical studies on the supposed "phenomenon," it is still commonly cited by debunkers like James Oberg and Phil "Dr. Phil" Plait to give an academic air to their knee-jerk dismissal of the Cydonia anomalies. Some of these debunkers even resort to claiming that articles written about "pareidolia" by other debunkers are some sort of paper trail proving the phenomenon has a publishing pedigree. But the simple fact is no such human tendency exists. At all.

There is however another very real human tendency that, unlike the mythical "pareidolia," is actually an extremely well-documented and medically established disorder—Prosopagnosia. Simply put, Prosopagnosia is a brain disorder that renders the poor souls that have it completely unable to recognize faces when they see them. According to some medical studies, as much as 2.5% of the human population may suffer from this disorder, and apparently a disproportionate number of those so afflicted have found jobs in the NASA planetary science community.

So the next time some NASA loving troll tries to tell you the Face on Mars is just all in your head, ask him to show you one medical paper—even one—which has studied the supposed phenomenon of "pareidolia." Then hit them back with Prosopagnosia. Within the next sentence or two I guarantee you they will call you a 'conspiracy theorist' or cite their academic credentials.

Since that writing, some of my more obsessive critics have sent me emails indicating that there is a reference to "pareidolia" in a medical journal from 1867. In reality, it is nothing but a single word in a 600 page document from 1867 and contains no information about any medical studies which establish the existence of pareidolia, since there never have been any. In any event, if this reference is genuine, it changes nothing as the term lay dormant for over a century before Goldstein used it to put the word into the modern vernacular. It is also clear from reading the two descriptions that the word may even have had a different meaning in the 1867 reference than it does as used by Goldstein the debunker.

They've also attempted to claim that three obscure papers mined on the internet use the word "pareidolia," in their abstracts, and that this somehow proves me wrong, and that pareidolia is a real, established medical condition. The three papers cited, which I was aware of when I wrote my original commentary, do nothing to make me change or retract my claim. None of them is a medical study that establishes that there is an actual medical condition which causes people to see things that aren't there or don't exist. All they do is make reference to the term.

The first paper, published in 2009, assumes that pareidolia" exists and specifically links it to facial recognition only. In reality, the paper is not about pareidolia at all, but only measures the speed of cognitive responses to visual stimuli, and nothing more. If the study had been done prior to Goldstein's 1994 reference and the debunking communities' repeated use of it with regards to the Face on Mars, the word probably wouldn't even appear in the abstract. In any event, the paper is not a study of pareidolia, but merely the speed of a particular brain function. The paper itself only uses the word pareidolia twice, and does not even cite a single medical or scientific study pointing to its existence.

The second paper, a tiny Japanese study published in 2012, tests patients with a specific kind of dementia—who already hallucinate—for hallucinations. It again uses the word pareidolia as a substitute for hallucination, but likewise cannot cite a single study verifying the existence of such a condition. In essence all it does is substitute the word "pareidolia" for "hallucination."

The third paper, published in 2009 in Brazil, is available only in abstract form and like the other two, assumes that pareidolia exists, but once again does not cite even a single study confirming that.

It's obvious that the critics simply went to the PubMed.gov website and searched for the term "pareidolia." All they could find were a pathetic three results, none of which is a study of the actual alleged "phenomenon." There is a fourth result, but the paper is in Spanish and the abstract is so off the wall that they must have decided not to cite it for fear of embarrassment.

What this reinforces is that as I have asserted, there is no medically established disorder called "pareidolia." If there were, there would be dozens of studies pointing to it and identifying its origins, causes, and treatment, including the portion of the brain that was operating in a defective way. Legitimate medical/psychological conditions like Prosopagnosia have such medical pedigrees. In fact, Prosopagnosia turns up an impressive 686 papers and studies from the same web site. The fact that all of the "pareidolia" papers were written *after* Steven Goldstein coined (or re-coined) the phrase in 1994 indicates they wouldn't even be using the term if it hadn't found its way into the debunkers' vernacular in the first place.

So all they have done in reality is to further prove my original point: the supposed phenomenon of pareidolia"is an urban myth perpetrated if not created by the debunkers, and nothing more.

So where does that leave the pictogram researchers? Are their studies legitimate or are they truly just "seeing things?" My take is that although these pictograms are likely not what the discoverers think they are, they remain a legitimate and worthy line of inquiry in Mars anomaly research. My reasoning here can be justified in three words—the Nazca Lines.

Located about 130 miles inland from the Paracas candelabra lay the Nazca plains. Although they postdate the candelabra formation by hundreds of years (scholars believe they were created by the Nazca culture sometime between 400 and 650 AD), the series of ancient geoglyphs are in many ways vastly more sophisticated. Laid out on a high, arid plateau, they are engraved in the hard, rocky

Portion of the "Nazca Lines" in Peru.

142

surface over an area of about 50 miles between the settlements of Nazca and Palpa, several hundred miles south of the Peruvian capital of Lima. Designated as a UNESCO World Heritage Site in 1994, the Nazca lines where brought into the popular culture by Ancient Astronaut theorist Erich von Däniken in his 1968 book *Chariots of the Gods?* In that volume, which sold millions of copies worldwide, and the later TV special *In Search of Ancient Astronauts* (1973), von Däniken pointed out the simple fact that the Nazca glyphs look remarkably like a modern day airport, with runways, interconnecting taxiways and even central "nodes" which might have once been terminals of a kind. He also noted that the formations are for the most part clearly made to be seen from the air, and that the primitive Nazca people were obviously incapable of flight. Von Däniken concluded that the Nazca Lines were in essence the signatures of a "cargo cult." He asserted that the Nazca plain may have once been a landing site for ancient astronauts and their flying vehicles, and that the Nazca people may have recreated the "landing strips" in an effort to lure the Gods of ancient times to

Parallel "landing strips" on the high plain of Nazca.

143

return. This is not unprecedented in the annals of anthropology.

In looking at the lines themselves, there is little to argue his point. The Lines do look remarkably like a modern day airport. At least one of the nearby mountain tops which contains glyphs and lines reminiscent of an airport landing strip appears to have been sheared off, almost as if by a cutting laser. Today we might be able to achieve something similar, but it would require tons of dynamite, which the Nazca didn't have, and the hauling away of massive boulders and debris, of which there is no evidence anywhere near Nazca. Certainly, such a feat of engineering was well beyond the primitive Nazca culture, whether it involved dynamite or death rays.

Sheared off mountaintop at Nazca, Peru. There is no conventional explanation for how the primitive Nazca culture could have flattened the mountaintop.

There are also hundreds of shapes and formations there besides the simple straight lines and geometric glyphs. Much like the alleged Martian pictographs, they represent all manner of earthbound life forms. More than 70 of the designs are stylized animals, plants and insects, including hummingbirds, spiders, monkeys, fish, sharks, whales, llamas, jaguars, a monkey and a lizard. There are also numerous plants and trees depicted. Scholars in general assign a religious significance to these formations, but cannot and do not provide any evidence for this beyond their vague speculations:

The geometric ones could indicate the flow of water or be connected to rituals to summon water. The spiders, birds, and plants could be fertility symbols. Other possible explanations include irrigation schemes or giant astronomical calendars.[1]

I've never heard of a spider being a "fertility symbol," since the female typically eats the male after coitus and is herself consumed by the babies when they hatch. The reality is, the conventional explanations of the meaning of the Nazca glyphs are pure speculation, which you can easily discern from the halting language in which the ideas are presented. This of course does not stop the academics and critics from attacking Erich von Däniken's ancient astronaut theories, although if the research were intellectually honest, his ideas would be given equal weight.

The critics go after von Däniken on several counts. First, they misrepresent his argument by claiming that he says the lines were made by the ancient astronauts themselves. In fact, he always argued that the glyphs were a recreation of something the Nazca people had witnessed previously, not that they were actually made by aliens. They also point out that there is nothing exceptional about them in monumental terms, and that contrary to popular belief, there are several vantage points in the area from which the glyphs can be observed from hilltops. While it is true that a very small percentage of the lines and glyphs can be seen from certain locations, most notably the nearby village of Cahuachi, the vast majority can only be discerned from the air. Even the *Wikipedia* entry for the Lines admits that "although some of the figures can be worked out from the surrounding foothills, the full designs cannot be truly appreciated unless viewed from the sky."[2] Even those who claim to know how the lines were made have admitted that they could not have been laid out without some kind of manned flight capability. Author Jim Woodmann postulated that the Nazca developed a primitive form of hot air balloon technology, but no evidence to support this has ever been found.

Exactly why and how a primitive culture like the Nazca would draw miles upon miles of lines and figures which can only be seen from the air is never really explained by the critics. The simple fact is that von Däniken's ancient astronaut cargo cult theory is probably the best explanation or at the very least on a par with

The "spaceman" at Nazca.

the conventional ideas. In fact, I'd go so far as to state that von Däniken's theory that the Lines are a form of communication meant for celestial beings is a far more supportable concept than any of the conventional models.

High on a mountaintop overlooking the central Nazca plain is a giant figure carved into the hillside. Alternately dubbed "the giant" by the conventional scholars and the "spaceman" or "astronaut" by Ancient Alien theorists, the figure is a humanoid form that seems to have either an exceptionally big head with very large eyes or to be wearing some kind of space helmet. Either way, it has one arm raised in a waving/greeting gesture, and I don't see how any rational person can conclude this is anything but a "land here" sign directed upwards to the heavens. It appears to be nothing less than an invitation for someone not resembling a human to drop in for a visit. The conventional scholars have no real explanation for why an ancient culture that they assert was primarily concerned with finding water sources would carve such a figure into a mountainside so it could be seen from the sky.

And the "celestial motif" doesn't stop there.

In the 1940s, Maria Reiche, a German mathematician and archaeologist, was recruited by one of the earliest Nazca scholars, Long Island University historian Paul Kosok, to help him determine the purpose and meaning of the Nazca Lines. They proposed the Lines were astronomical markers on the horizon to show where celestial bodies rose and set, although why a primitive people like the Nazca would be interested in such celestial mechanics was not considered. Later, Reiche proposed that some or all of the figures represented known constellations or at least celestial diagrams. In 1998, Phyllis B. Pitluga, a protégé of Reiche and senior astronomer at the Adler Planetarium in Chicago, concluded that the animal figures were "representations of heavenly shapes." She performed a computer analysis of star alignments and asserted that the giant spider figure was an anamorphic diagram of the constellation Orion. She further suggested that three of the straight lines leading to the figure were used to track the declination shift of the three stars of Orion's Belt.[3, 4]

This of course got her into immediate hot water with the anti-ancient astronaut crowd. Dr. Anthony F. Aveni quickly jumped down her throat with a scathing critique: *"I really had trouble finding good evidence to back up what she contended. Pitluga never laid out the criteria for selecting the lines she chose to measure, nor did she pay much attention to the archaeological data Clarkson and Silverman had unearthed."*[5]

Of course, despite the criticism from academia, it is quite

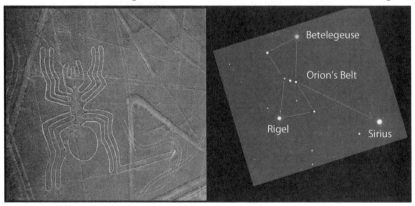

The spider figure as a mirror image of Orion and Sirius.

147

obvious that Pitluga is correct, with a slight modification.

While I agree that the spider figure is a representation of Orion, one aspect of the drawing that is not discussed is the very long leg stretching out 90 degrees to the right. I recognized this as an aspect of what author Graham Hancock calls "Heaven's Mirror," a Peruvian version of the Egyptian axiom "as above, so below." If you simply take an image of the constellation Orion and the star Sirius—Osiris and his sister/consort Isis to the Egyptians—and flip them to reflect a mirrored orientation, they align almost perfectly with the spider figure. The long leg extension is perfectly proportioned to indicate the position of Sirius relative to Orion in the night sky.

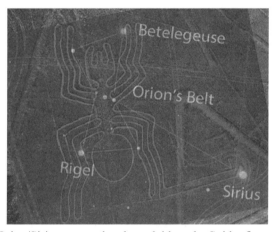

Orion/Sirius reversed and overlaid on the Spider figure.

Other ancient cultures, especially the Egyptians, recognized and reflected this sacred arrangement and relationship between these two celestial bodies in their monumental architecture, in the form of the great pyramids at Giza. That the Nazca, millennia later and in a far more primitive manner would acknowledge the same key celestial arrangement is compelling, considering that the two cultures had no contact. What was so important about these two figures that virtually every ancient culture was fascinated by them?

While that question remains a mystery, there is no doubt that the Nazca were far from a simple agricultural people. They were undeniably fascinated with every aspect of the movements and meanings of the stars, and in the spirit of "as above, so below," they carved out these figures as an homage to either the heavens themselves, or perhaps the beings that they had previously met who came from the heavens. It therefore follows that if the glyphs were laid out to connect the Nazca people with the stars, then the Lines, with their runway-like appearance, were most probably meant to represent the same thing. They were both meant to connect, or reconnect, as von Däniken guessed, with the Star People.

What then does this say about the alleged pictograms of Mars? Are they the same thing, a message to the people of Earth who might find them indicating that we are not alone amongst the stars and never have been? Or are they truly a figment of the viewer's imagination, a set of "Nazca Lines" only in the mind of the observer? To address that question, I decided to go back once again to that most enduring of Martian mysteries, the Face on Mars itself.

As far back as his 1992 United Nations speech, Richard Hoagland had proposed the possibility that the Face might have actually been meant to be seen as two faces, one human and one feline, split down the middle. Originally inspired by the fact that the right side of the Face was significantly asymmetrical when compared to the left in the original *Viking* data, Hoagland decided to run a series of photographic symmetries of the two sides of the Face. The results were interesting, if not compelling.

Most people's immediate impression from these symmetry studies was that the left side of the face looked distinctly human, or at least primitive hominid (Hoagland freely admitted "we're not talking Paul Newman or Marilyn Monroe here"). The other side, the

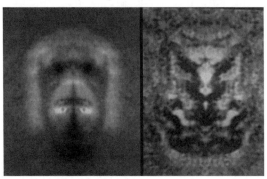

Symmetry views of the left and right side of the Face on Mars from *Viking* data.

149

less distinct, more shadowed side in the *Viking* data, was decidedly feline. It looked like a lion.

This raised all sorts of new questions about the Mars/Egypt relationship, and whether the Face was intended as some sort of "Martian Sphinx." Contrary to what you might expect if this was simply an optical illusion caused by the lower resolution of the *Viking* data, the impression was only enhanced by later, higher resolution *Mars Global Surveyor* images of the Face.

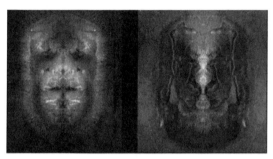

Mars Global Surveyor symmetry views of the Face on Mars.

So with the "trick of light and shadow" argument now permanently disproved by the MGS and later images obtained from other instruments, the question became one of meaning. What was in the mind of the theoretical Martian architect? And did it persist across other datasets?

Did it ever.

The direct overhead views obtained by *Mars Express* held some interesting surprises when they were subjected to the same technique. They showed that not only did Face/Lion sides retain their appearance, but it also revealed a series of bizarre, geometric patterns both above and below the Face mesa itself.

These patterns look exactly like a decorative headdress above, and a strange set of alien skull-like faces below. How these could be visible in the *Mars Express* data but not really show in other datasets can simply be explained by the fact that all CCD cameras are different. Just as the THEMIS instrument "sees" details that are not visible in the MRO images for instance, likewise all CCD

Geometric patterns above and below the Face on Mars.

cameras are going to see details differently in visible light.

This is simply because each instrument is made up of different components, with different strengths and sensitivities, whether it's the *Viking* Vidicon system, the MGS "Malin Camera," MRO's HiRISE camera, or ESA's HRSC camera. They all have different lenses, different sensor grids, different internal hardware, and are programmed to operate differently. So in this sense it's not

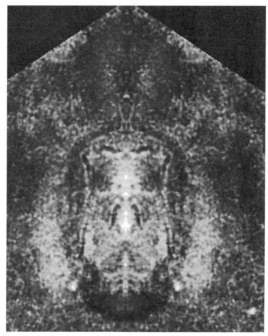

Close–up of the "Headdress" from right-hand symmetry of the Face on Mars.

surprising that the ESA images show details the others do not. What is extraordinary is *what* they show.

The "headdress" is particularly interesting to me because the pattern is so geometric and non-fractal. How the ground around the

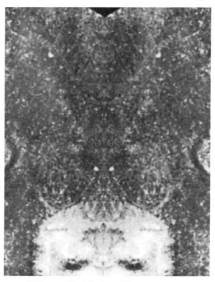

"Human" side headdress pattern.

152

Face could have this organized pattern is a mystery to me, but I have no doubt it is really "there" because I can show clearly that they are not imaging artifacts and since they appear ever so faintly in other datasets of the same area. Since the original source data are uncompressed TIFFs, there is no possibility they are JPEG compression artifacts. The other thing that I really have a hard time getting my head around is that some of this geometry actually seems to be *suspended above* the Face itself, rather than "imprinted" on or around it.

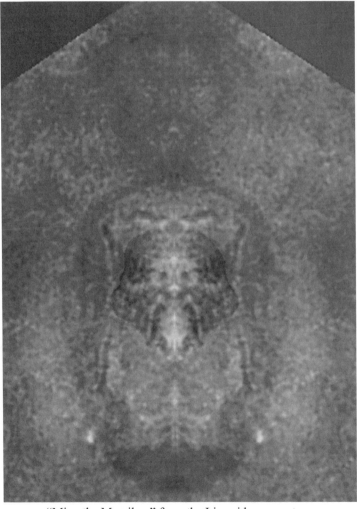

"Ming the Merciless" from the Lion side symmetry.

But that was nothing compared to how weirded out I got when I saw "Ming the Merciless."

As I was studying the images, my eyes were drawn to very human looking face within a Face on the upper part of the Lion (right) half symmetry. The eyes, nose, ears and a Fu-Manchu style mustache really jumped out at me from the image. But it raised the question: what the hell was it *doing* there? Had I really fallen down the rabbit hole of pareidolia, or was it actually there, embedded in the data?

And that raised more questions. Was this image reflective of what the Face really looked like? Had the Martians actually put these shapes there to attract our attention? Or was somebody at ESA simply having some fun at our expense with Photoshop? I truly had no answer.

If this was actually an intended visual message, then that had all sorts of implications that I simply could not answer. Among them were, had humans or something very much like us, actually walked the surface of Mars in the recent past and built this monument knowing we would study it in this way? There certainly is a cultural precedent among terrestrial cultures, with half-human/half-feline ceremonial masks discovered among Inca and Mayan relics.

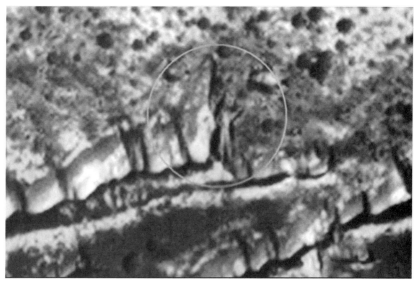

MRO image ESP_011359_1695_RED.[6]

154

But it goes even deeper and gets even weirder when we consider discoveries on other parts of Mars.

While attending a conference in Los Angeles, I received an enigmatic image taken from *Mars Reconnaissance Orbiter* that was given to me by researcher Malcom Scott. Taken from a small section of MRO image ESP_011359_1695_RED, it shows a shape amongst what appear to be layers of water eroded cliffsides. That shape, to everyone who's ever looked at it, is unquestionably an upright, bipedal… cat.

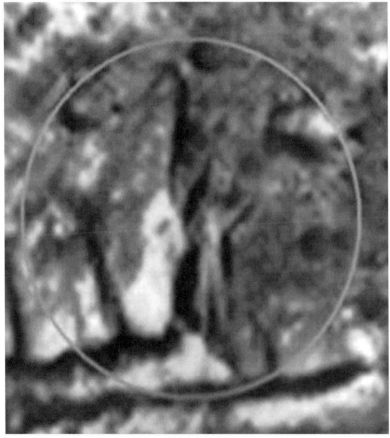

The "air guitar" cat on Mars.

As bizarre as that sounds, the shape seems to be undeniable. This cat, which one observer noted seems to be playing "air guitar," is standing at the entrance to a dark cave or at least an alcove which is shaped to contain it. It has two legs, two arms, a torso, shoulders, a long neck and a distinctly feline head. It looks exactly like the kind of statues that the Egyptians used to place outside their temples and monuments. Having said that, the area around it looks like a product of natural erosion, and there is not much else in the area that might be artificial. This raises the question of just what a single, air-guitar-playing cat statue would be doing on Mars, in this particular location and in complete isolation. If this cat guy appeared in Cydonia amongst the ruins, I would be absolutely down with calling it artificial. But given its isolation, I have a really hard time doing that.

What I ultimately decided was that I would never be able to answer the "why" of these observations. All I could do was attempt to understand if the observations were valid. Or, see if these "pictographs," statues and symmetry studies led me in another, perhaps even more fruitful direction.

As it turned out, that's exactly what happened.

(Endnotes)

1 Brown, Cynthia Stokes (2007). *Big History*. New York: The New Press. p. 167. ISBN 978-1-59558-196-9.

2 http://en.wikipedia.org/wiki/Nazca_Lines

3 http://www.zoominfo.com/p/Phillis-Pitluga/1158172653

4 http://en.wikipedia.org/wiki/Nazca_Lines

5 Aveni, Anthony F. *Between the Lines: The Mystery of the Giant Ground Drawings of Ancient Nasca, Peru*. Austin, Texas: University of Texas Press. p.205. 1 July 2006. ISBN 0-292-70496-8.

6 http://hirise.lpl.arizona.edu/ESP_011359_1695

Chapter 7
Parrot City and the Eye of Ra

Since the early days of the Cydonia investigations, the number of independent Mars researchers has skyrocketed. This has enabled so many more "eyes" to be focused on Mars that the number of discoveries has skyrocketed along with it. The pictogram researchers were among the earliest in the age of the internet, and while I can't really endorse most of the claims they've made, their observations have occasionally (and inadvertently, on their part) led me to discover the most extraordinary proof of my core thesis. A case in point: "Parrot City."

First noted by pictogrammist Wil Faust on MOC narrow-angle image M14-02185, the "Parrot" is a formation that vaguely resembles the terrestrial avian. Using Photoshop, researchers like George J. Haas of the Cydonia Institute and Joseph P. Skipper of the Mars Anomaly Research site[1] have argued that the Parrot is a sculptured formation made to purposefully evoke the terrestrial bird. Haas even co-authored a paper on the subject[2] and made it a focus of his book *The Martian Codex*.

The "Parrot" is located in the West Argyre basin, a 1,100 kilometer wide, deep crater-like depression in the southern hemisphere of Mars. Located between 35° and 61° South and 27° and 62° West, the Argyre basin and the Parrot region has been imaged frequently by various NASA probes. There are no less than three specific targets by the *Mars Global Surveyor* spacecraft of this area: M14-02185 in April of 2000, S20-00165 in July of 2006, and S13-01480 in December of 2005 (the "S" stands for "special," meaning the latter two images were singled out as specific targets of interest). In looking at the images, it's easy for me to see why NASA would be so interested in this area, but probably not for the reasons that the pictogramists believe.

The "Argyre parrot" from M14-02185, with shading and colorization added to emphasize the form.

While the "Parrot" formation does bear a passing resemblance to the 50 million year old terrestrial avian, it takes quite a bit of shading and emphasis to make it stand out. As noted by Faust, just north of the formation is a grid-like set of city structures he spotted and nicknamed "Parrottopia." It is this "lakefront property" that I strongly suspect was NASA's real interest in the area.

When I first saw Parrot City I felt it was the most obviously artificial set of structures I had ever seen on Mars. It is laid out exactly like a lakefront town here on Earth, with roads, bridges, block-like buildings and what appear to be industrial areas as well. There are no less than five parallel roads that run over land bridges through the mountains just north of the "Parrot" right into the heart of the city. They are intersected by other clear cut roads which run at 90 degree angles, forming cul-de-sacs and neighborhoods for what look like individual single-family structures. There are even "piers" at the end of these roads which stretch out into what is now a dried up lake bed or river. Along with that are some unmistakable features of advanced (but ruined) inhabitation, including vertical structures, retaining walls and complex tubular shapes.

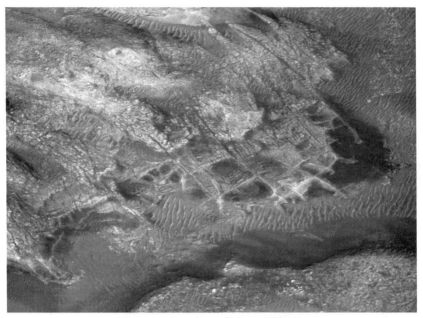

Perspective view of "Parrot City."

The wavy dunes in the low-lying areas are what used to be the bottom of the river/lake that this city used to reside on the edge of. Simply darkening in the low-lying areas shows where the water once flowed, with the wave action churning up the bottom and leaving the fossil signatures of the "dunes."

The image strips are about three kilometers (almost two miles) wide, the "Parrot" formation about one and half miles across, and the lakeside town is about the same.

The area looks not entirely unlike the affluent Redondo/ Hermosa/Manhattan Beach areas of southern California with their jutting piers, rectilinear access roads and blocks of residences. In fact, the scale, layout and organization are virtually identical. This becomes even more obvious when you compare them side by side (or top to bottom, as the case may be).

159

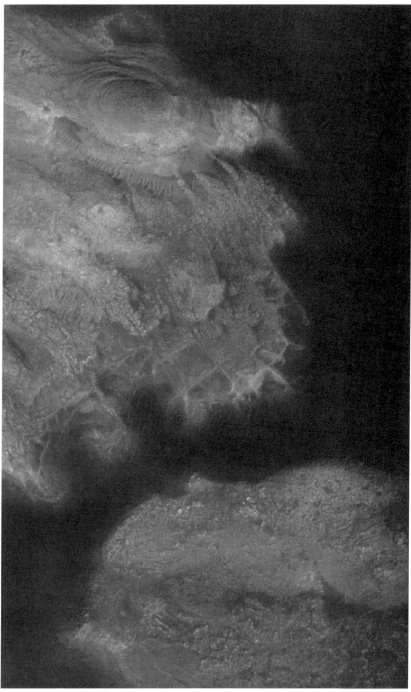

Map projected version of the area around the "Parrot" and the lakeside city. Low-lying areas have been darkened to show where water once flowed.

Hermosa/Argyre comparison.

But as we get into the details of the Argyre Parrot City, we can see even more evidence of intelligent construction. In fact, there seems to be quite a number of unusual objects in and around the Parrot City area.

High resolution images of the area around the Parrot City show a partially buried but extensive crisscross pattern just beneath the surface dust. Such patterns are usually indicative of intelligent design. This particular pattern is occasionally disturbed by objects like the "waffle iron," an immense block-like structure sticking hundreds of feet above the flat surface of Mars. It has four distinct and square cells inside it. Quite simply, there is no natural geologic explanation for its origin. That said, I have no idea what it is or what its original purpose was. It may just be a tipped over building. But what I do know is that it's artificial. And there seems to be a great deal of artificial debris around it.

As I investigated the area further, I next addressed the "Parrot" itself. While Haas and others had gone to a great deal of effort to support their hypothesis, in studying the "Parrot," I saw no such

The "waffle iron" near the Argyre Parrot City.

evidence of artificial design. The pictographers were quite good at pointing out various features that compared anatomically to a terrestrial parrot, but in the end I was unconvinced. Most of what I saw on the "Parrot" formation looked to me like natural water or wind erosion, which fit with my hypothesis that both the "Parrot" and the town were once bordering a lake or riverbed. The "body" of the Argyre Parrot particularly shows the swirling, layered signatures of typical fluvial erosion processes and there is no evidence of the type of geometric signs of construction usually seen on other Martian monuments. It also has some geologic features typical of windblown formations called yardangs.

So, satisfied that the "Parrot" was the gateway to the real story rather than the story, I turned my attention to the details of "Parrot City." It was there that I would find either the truth or the fallacy of my longer distance observations.

Fortunately, between the three datasets (and NASA doesn't waste precious orbiter resources photographing the same thing three times unless they are *very* interested in it) it was a fairly straightforward task to get down into the details of the images and

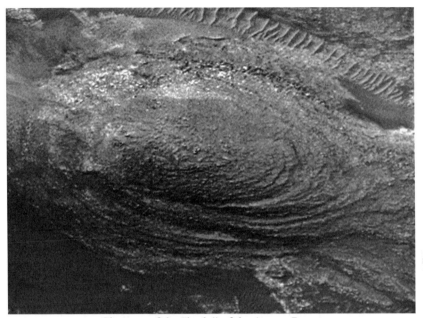

Close-up of the "body" of the Argyre Parrot.

identify individual structures. The best of the three images is an uncompressed, 14MB TIFF which has a spatial resolution of about 1.5 MPP or meters-per-pixel, good enough to spot objects as small as large semi trucks. I decided to break my detailed analysis up into four specific sections of the City, starting from the right and working my way to the left. I nicknamed the area to the right "The Pier" because it contained several roads that extended pretty far into the darkened areas, which I thought comprised the dried up lake or riverbed. If you look at the overall view of the entirety of Parrot City, there are actually at least five (if not six) such parallel access roads running from the Parrot formation through the mountains and down into the city. The most prominent road runs down into the dark area where intersecting crossroads or bridges meet it at a 90 degree angle. It then extends into the riverbed where it bends downward and terminates in what looks to be a square piling embedded into the river bottom.

The "bridges" are especially obvious artificial structures, connecting to the main road as I stated at 90 degree angles and in some cases intersecting with other access roads. The ground in

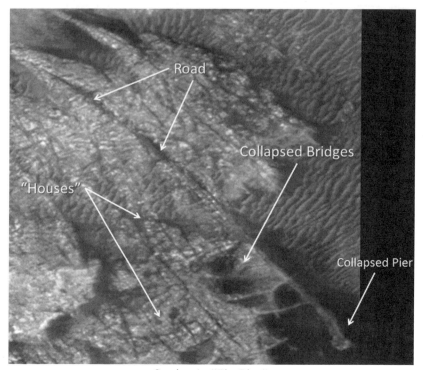

Section 1– "The Pier."

Collapsed areas around the bridges showing underlying supports and collapsed "pier" anchored in the former lakebed.

between and around the bridges seems to have either collapsed or been somewhat filled in by debris flow. The support beams for the bridges are clearly visible below the spans. Above the bridges are a series of block-like individual structures that are sized exactly like single-family homes. More such structures are visible to the left side of the image just across from another of the parallel roads I mentioned earlier.

But in the lower part of the image is a squared off area about the size of a city block. I call this the "Rotunda" because there is a rounded central structure about the size of a two-story hotel with a parking lot. Just below the rotunda are exposed geometric cellular structures which probably support the street above. One could easily drive a car up, enter through the driveway, park the vehicle and then walk to the pier.

Exactly like people do right here at the beach cities of California.

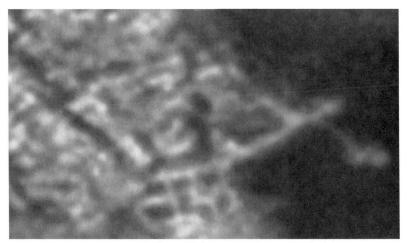

Close-up of the "Rotunda."

In short, the pier area of Parrot City looks exactly like any beachfront community on Earth. The specific function and level of detail in the images rules out any possibility of a natural geologic explanation. But the pier section of the city is by no means the most obvious or impressive area in these images. That honor is reserved for the "Industrial Complex."

Section 2–The "Industrial Complex."

Immediately adjacent to the pier area and covering multiple city blocks in size is an area I call the "Industrial Complex." It contains all kinds of anomalies which can't possibly—under any scenario—be products of natural erosion. While I can only speculate as to their actual function, there is no doubt that the area looks exactly like industrial areas of large cities on Earth, and at the same scale. There

The "Processing Plant." Entrance to the storage hangar is visible in upper right.

are even more houses in organized neighborhoods situated around it. For identification purposes, I have nicknamed the main features the "Processing Plant," the "Storage Hangar," "Waste Disposal," the "Office Park," the "Housing Development," the "Lighthouse" and the "LAX Restaurant." Each of them in isolation is extraordinary. Viewed as a collective assemblage of structures, they are quite simply beyond natural explanation.

The processing plant is inside a squared area about the size of a typical city block here on Earth. To the northeast (upper right) of the structure is a low, bunker like housing which resembles an airplane hangar or storage facility. There is even a darkened road leading from the processing plant to the bunker. The processing plant has an enormous main entrance, accessible from several driveways that lead into the flat "parking lot" outside the entrance. On the roof just above the entrance is a square, four-sectioned structure that looks vaguely like a vent or some kind of solar panel assembly. Its tubular frame is connected to the ground by several pipes or conduits. Toward

The "waste disposal" facility. Dark area at the top is the entrance to the processing plant.

167

the center of the roof are some more pipeline-type structures, one of which leads to a starfish shaped structure and another of which leads to another crosshatched, tube-framed "solar panel" on the roof in the upper left corner. Below that at ground level are three archways which look as though they might be entrances for large trucks or people. The "hangar" appears to be built on a berm slightly elevated from the main structure. While I have no idea what the purpose of this facility is, I have no doubt that it IS a facility, and there are several other nearby structures that appear to be linked to it both by surface roads and presumably by underground tunnels.

Just below the processing plant, the hillside slopes down dramatically. In a squared-out area just below the entrance to the processing planet there are again a series of block-like "housing" structures and parallel ramps that run down the hillside. One of them leads to a flat, bright, jagged edged and highly reflective platform that runs for several hundred feet across the middle of the frame. This sculptured terrace is positioned right above a series of vents which jut straight out from the sloping, angled hillside. Below the vents is another flat platform, and below that a large open area with a flat floor that I call the "loading dock." There is also what appears to be tons of debris to the far left, and perhaps some kind of massive

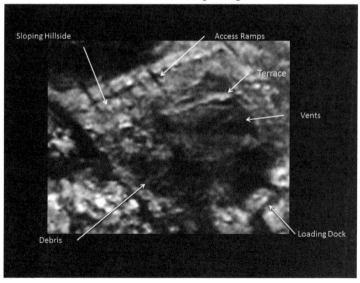

Labeled view of the waste disposal area.

pump mechanism.

Again, because of its proximity to the processing plant and its position adjacent to it, I chose to label this area the "waste disposal" section of the industrial complex. Whatever might have been mined/processed/manufactured there would obviously need some kind of mechanism to vent gasses or dump liquids, so if this area were in fact an industrial complex this is exactly the kind of support facility we would expect to see. The "loading dock" almost looks like a helicopter pad from which supplies could be brought into the underground section of the processing plant.

The "Office Park."

Of course, no industrial complex would be complete without an accompanying office park nearby from which the engineers could oversee the operations. Such a facility would be connected to the plant through a series of underground connecting pedestrian tunnels, exterior walkways or ramps, or perhaps some kind of tram system. Immediately across the street and adjacent to the processing plant is just such an office park. This area is a terraced, multi-layered facility which connects directly to the processing plant. The front of the office park has darkened "windows" with clearly visible vertical struts supporting the various tiers of the facility. Above there is

Terrestrial comparison, Federal Way, Washington.

even a series of bunkers carved into the hillside. Various external structures on the "roof" of the office park give a more industrial, functional appearance.

Such facilities are hardly unprecedented here on Earth. The Weyerhaeuser international headquarters in Federal Way, Washington is one example that is laid out in a similar fashion. It has the same layered, terraced look, although its terraces are covered with plants to give it a more nature friendly aesthetic. However, the same equally-spaced vertical posts are visible, and on the roof we can see various air conditioners, pumping houses and other industrial support structures. To me, the entire Parrot City office complex is fascinating because of its adherence to the forms, fits and functions of terrestrial architechture.

Of course you can't have processing plants, waste treament facilities or office parks without workers to man such facilities. Fortunatley Parrot City seems to have the best in such housing for its workers as well.

As already described, there are various places around Parrot City which have regularly distributed "blocks" of what look like individual housing structures. To the far left of the industrial

The suburbs in Parrot City near the industrial complex.

complex section of the city we can see further examples of this. Arranged in small neighborhoods of eight to 10 structures, we can see interconnecting roads, cross streets and even what may be chimneys or towers on some of them. There are even obvious driveways and garage entrances. And all of these are within easy walking distance of the office park and processing plants.

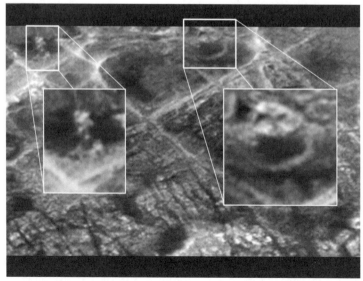

The "Lighthouse" and the "LAX Restaurant" on Mars.

171

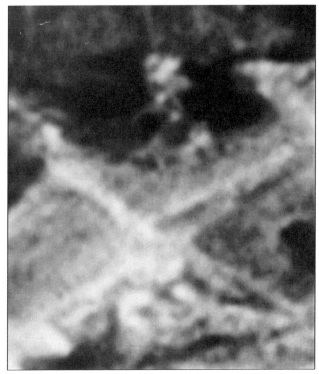

Close-up of the "Lighthouse."

There are, however, a couple of even weirder and more obviously artificial structures visible in the Industrial Complex area of the city. I defy anyone to explain either of them by any natural geologic process. I call them the "Lighthouse" and the "LAX Restaurant."

Of the two, I find the lighthouse the most intriguing because it is so mysterious. "Lighthouse" is merely a nickname for a vertical structure that I simply cannot identify. It is situated on the edge of where the town transitions to the dried up lakebed, and it sits in the center of a city block defined by three distinct access roads at 90 degree angles to one another.

The structure has a bright white base with regularly spaced darkened areas which may be ground level entrances to the main structure. On the roof of the base level is a dark, bulbous protrusion that might theoretically house a large mechanism or motor for moving the main antenna. It looks a bit like a nuclear reactor

The "LAX Restaurant"—on Mars and Earth

containment building, although by no means am I suggesting that's what it is. The multi-bulbed main "antenna" has five distinct nodes and a beefy vertical mast extending from the base up into the antenna assembly. At the scale of the image the "Lighthouse" has a height of at least a couple of hundred feet, making it the tallest structure for some considerable distance around the area. What its function might be I can only speculate, but it has no signatures of a naturally occurring object, and numerous indicators of a constructed object.

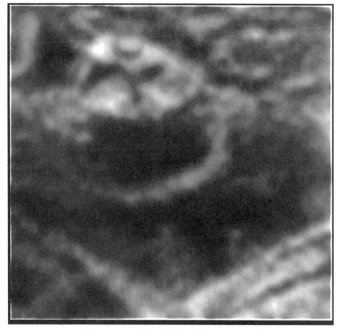

Close-up of the "LAX Restaurant."

173

The "LAX Restaurant" sits on a hill just above the processing plant. Like its namesake, it has three (or four) distinct, futuristic arches that are equally spaced around a central raised dome. The base of the structure is a brightly colored disk, and there are hints of something in the shadowed area beneath the arches—perhaps the entrance where you take the elevator up to the restaurant for your meal with a view.

Below the platform upon which the restaurant structure resides is a dark, cylindrical structure which might be a liquid storage tank or some other kind of storage unit. To the left are some brighter, block-like objects which may at one time have been "casing stones" of some type. Below the storage tank is a second bright-white disk shaped platform parallel to the upper one. Again, these sorts of features simply don't occur in nature, but rather only in constructed facilities with an artificial origin.

As bizarre (and bizarrely artificial) as these two objects were, they were quickly forgotten when I got a good look at the next two areas, the "Race Track Complex" and the "Pumping Station."

The race track complex is adjacent to the industrial complex and originally caught my eye because of the two dark oval shaped trenches dug into the flat area encompassing, once again, about a

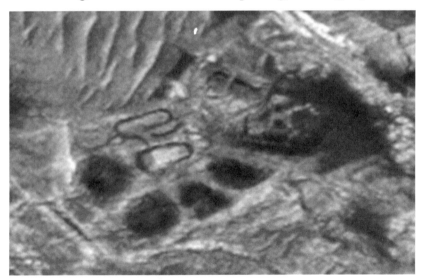

Close-up of the "Race Track Complex."

174

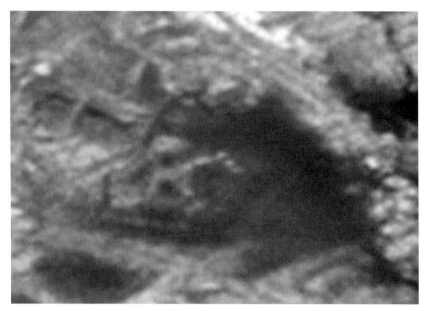

The train platform.

city block in size. Closer observation showed that the two were connected either by a road or an overpass, and there were four other dark depressions configured in a pattern around the "tracks." These almost had the look of reservoirs, and while I'm not saying that's what they are, they certainly could be some kind of holding tanks or storage areas. There's even a hint of a vent or outflow tube in the lower right corner.

These two oval shaped "race tracks" and the equally sized and spaced "reservoirs" around them would be anomalous enough, but just above and beyond them, built right into the hillside, are some equally curious if not outright artificial structures. To the left and built into the hillside much like the office park there are what look to be bunker windows, with a central spar reinforcing the structure vertically. Immediately to the right of those is a massive blockhouse, housing who knows what kind of mechanism. The blockhouse sits atop a square platform, which has hoses or tubes connecting to it from above; the tubes look to connect to what looks like—let's face it—a school bus. The "school bus" may simply be some kind of transport truck, collecting ore, oil or some kind of liquid for distribution to

175

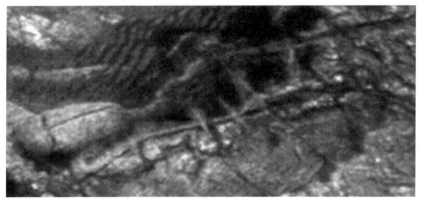

The Pumping Station.

somewhere else. The blockhouse may very well be an entrance to some kind of mine, and it's possible it may slide under the mountain over it, drill out some ore of some kind, offload that to trucks and then ship it to the processing plant for, well, processing. We will only know if we ever land there and explore. But the whole concept is supported by the next area to examine in detail—the "Pumping Station."

Situated just above the race track complex and overhanging what may be the mine entrance is what I call the pumping station. It is uniquely distinguished by the rounded formation at the lip of the race track complex which some have likened to a scarab beetle but which I nicknamed the "potato bug." Such a large round formation is unlikely to made by nature, especially in this context. It almost looks like an inflated balloon, held to the ground by a series of tethers and supports. But it is clear when looking at it more closely that there is a series of pipelines coming out from the mountains and feeding into the object. One main flexible hose emerges from the back of the potato bug, and twists and wends its way to attach to a more rigid pipeline which runs off into the mountains. Just to the right of that pipeline and running in parallel is a second rigid looking pipeline which has cross pipes at 90 degree angles all along its length, some of which seem to connect to the first pipeline. At the "nose" of the potato bug there is also a smaller-diameter pipe that connects to the base which also has vertical supports to hold it up.

Although I have no doubt that this entire area is artificial, I cannot say exactly what it is used for beyond the speculations I have made here. But I will go to the mat with anyone who thinks they can explain it as anything other than a snapshot look at some long-abandoned Martian industry.

Individually, any one of these areas constitutes jaw dropping proof of artificiality. There is no question whatsoever in my mind that the entire Parrot City area was constructed long ago for who knows what purpose. But just when I was prepared to leave it at that and move on to the next subject, the Parrot City pulled me back in.

In the course of doing my due diligence on the Parrot City formation, I did some searching on it to try to find all the images. As it turned out, NASA/MSSS had taken a fourth image of the region at

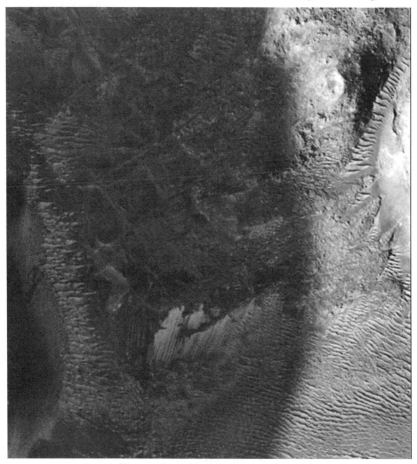

Parrot City as viewed by MGS image S20-00165p.

the request of an anomaly hunter per the public targeting initiative by MSSS. Released on August 11, 2006, the image is taken in much different lighting than in the other three images. The parrot formation pretty much disappears, but Parrot City beyond is seen as it has never been seen before. Draped in shadow, it provides a completely different perspective on the city. A perspective which only deepens the mystery and creates more questions.

Due to the shadows across the city, the details that we saw in the other images are lost in noise and deep shadow. But what stands out is the fact that what appeared to be dark, sandy lake bottom soil in the other images is in fact a depiction of bright, reflective and obviously artificial *vertical retaining walls*.

These retaining walls appeared in all the other images as virtually pitch-black smooth soils upon which the "pier" and nearby structures such as the rotunda rested, just a few dozen feet above the old lake bottom. This new perspective shows that they are in fact not dark at all, but bright (they have to be to show up so well in deep shadow) with parallel vertical grooves or spars that look exactly like structural rebar. This "rebar" is the supportive underpinning of not only the retaining walls, but maybe even *the plateau upon which the entire city rests as well*.

Close-up of Parrot City "retaining walls" as seen in MGS image S20-00165p.

The retaining walls themselves are hundreds of feet high. By my estimation the tallest of them may be as far as 500 to 600 feet above the lake bed below, instead of being virtually even with it, or simply showing a shallow drop-off as we saw in the previous images. What this means is that our previous perspective of the area was not correct. The far right "pier" is not at a parallel height with the rest of the Parrot City complex, but instead is perhaps 100 feet below the mean "city plane."

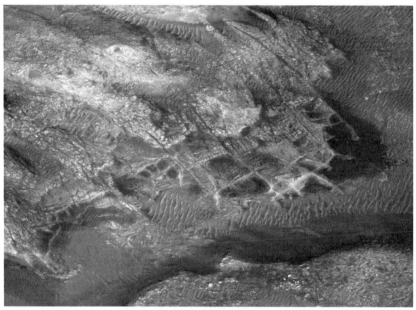

NASA/MGS image 2005_08_22_M14-02185.

This illusion was created because the original Parrot City data released by NASA/MSSS (NASA/MGS image 2005_08_22_ M1402185) was incorrectly map projected. In the "map projected" data presented by NASA, the entire image strip was stretched horizontally far beyond its correct proportions, resulting in an optical illusion that the retaining walls were flat, not vertical, and part of the lake/riverbed. Or, at best, they constituted a shallow drop-off. This assertion can be confirmed by simply compressing M14-02185 horizontally, until it matches the newer S20-00165p image.

NASA/MGS image 2005_08_22_M14-02185, compressed to match the perspective of S20-00165p.

Once this is done, we can see the first "pier" we examined actually rests on a retaining wall some distance below the others, but still hundreds of feet above the old lakebed. The cross bridges that we saw now appear as almost vertical or at least deeply angled "pilings" supporting the upper part of the City. A couple of the other piers look like bright edges of the retaining walls pointing more downward than horizontal, as we first observed. But there is a bigger problem here. The difference between NASA/MSSS images M14-02185 and S20-00165p also involves the so-called "non-map projected" versions of M14-02185 and the earlier data. Supposedly, they have not been ortho-rectified and therefore should not have this distortion of perspective. That is why I, for one, always choose to work with non-map projected data whenever possible, as I did in this case. Yet both the map projected and non-map projected data show the same problem. This means that it is a virtual certainty that even the non-map projected images were altered (stretched horizontally) by someone at NASA/MSSS in order to take away the

180

visual cues that would show the true vertical height of the retaining walls. Given that they would be a virtual give away that the area in

NASA/MGS images S20-00165p (top) and M14-02185 (bottom).

question had been artificially transformed by what must have been mind-boggling technologies, it's also easy to see why they followed up by blackening out the walls themselves, leaving only a hint of the vertical striations (rebar).

The fact is, there is simply no way walls bright enough to show up in an image taken when the area was in deep shadow can likewise appear pitch-black in the earlier images, when Parrot City was in full daylight. Even accounting for different sun angles, if the walls were bright enough to show up in S20-00165p, they were well more than bright enough to show up in M14-02185. The only way they could not is if someone at NASA/JPL/MSSS deliberately and with malice aforethought darkened them to the point where they disappeared.

I should think the reasons they did this would be evident. Such obvious retaining walls would be a proverbial smoking gun that the Parrot City area was artificial. Not even the most willfully ignorant of NASA's planetary scientists could credibly claim that the features were natural, especially when the rest of Parrot City is taken into account.

As to why they did not likewise blacken out the retaining walls on the newer image, S20-00165p, I can only speculate. However, I strongly suspect it has something to do with the fact that S20-00165p was specifically requested by a member of the public as opposed to being within the purview and control of the elite, i.e. Dr. Malin and his cohorts at NASA and JPL. But that is getting into *Dark Mission* territory and is beyond the scope of this book.

In any event, none of this changes my previous observations or conclusions about any of the data presented in this chapter. In fact, it rather dramatically increases the chances that *all* of these objects are artificial, and I was 100% certain of that before I found S20-00165p. But it also points out that the pictogram hunters, whatever you may think of their analysis and conclusions, provide a valuable service to the independent research community in general. If Wil Faust hadn't looked at M14-02185 and thought he saw a pictogram of a parrot, no one might ever have bothered to do a deep analysis of the area around it, and we would have missed one of the most profoundly significant finds in the history of Mars anomaly research—a mountain city akin to the mythical capital of Minas Tirith from the Lord of the Rings series. Maybe that's what the Lighthouse is after all, the white tree at the top of the mountain city…

Before we move on, I want to examine one more such mystery on the surface of Mars, one discovered by one of the most dogged and determined of the independent researchers, Gary Leggiere.

Calling himself the "Mad Martian," Gary, like Joseph Skipper and many others, has poured over images returned from the *Mars Global Surveyor*, *2001 Mars Odyssey* and *Mars Reconnaissance Orbiter* probes for years. He has focused mainly on searching for more faces like the Face on Mars at Cydonia, with mixed results. But recently, he spotted something using the "pictographic mentality" that has led to what I consider a major discovery—the "Eye of Horus."

The "Eye of Horus" from MRO image B19_016974_2190_XN_39N020W.

Captured by the MRO's context image camera in 2010, B19_016974_2190_XN_39N020W has a scaled pixel width of about six-MPP, far better than many primary visual image cameras on many of the probes sent to Mars in the last decade and a half. While it seems to be in isolation, it is just too weird to ignore on that basis.

Gary was immediately drawn to certain features oe the object that seemed out of place. While there were many knobs and buttes

183

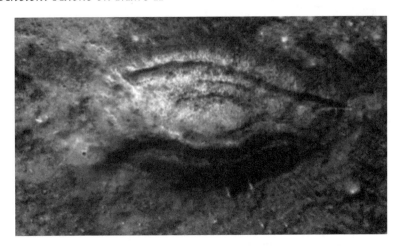

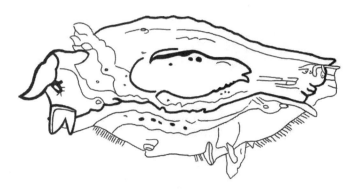

"Eye of Horus" and an avian comparison (Cydonia Institute).

in the image strip, none of them had the weird shape and partially buried character of the "Eye." He was attracted to the apparent eye shape and the fact that it seemed to have structural characteristics. It begged for further examination and he promptly posted it on several message boards, including that of George Haas' Cydonia Institute.

George felt that, as with the Parrot formation, this object had a distinctly avian shape and felt that it may have been intended as another Martian pictogramic sculpture. But there was another reader on the board named Steve Meads who had written a book titled *Chauvet Dreams* who had another take on what the object might be, and after several back and forth attempts to find out the details of it, he posted about it on his "Mystical Files" blog.[3]

Meads pointed out the long and ancient connection between supposedly primitive peoples like the ancient Egyptians and the planet Mars. He specifically connected the ancient Egyptian god Horus with Mars, as many others have noted before. As it was put by author Graham Hancock in his book *The Mars Mystery, the Secret Connection Between Earth and the Red Planet*:

> *The ancient Egyptians envisaged a profound connection between Mars and Earth and, more specifically, between Mars and the Great Sphinx of Giza. The planet and the monument were both seen as manifestations of Horus, the divine son of the star-gods Isis and Osiris. The planet and the monument were also both called by the same name, Horakhti, meaning "Horus in the Horizon." Mars in addition was sometimes known as Horus the Red, and the Great Sphinx, for much of its history, was painted red.*

Beyond the Great Sphinx, which has often been compared to the Face at Cydonia because of the human/lion aspects they share, this connection was represented by the Egyptians in their other sculptures, statues and heraldry. The specific pictogram of Horus, the Mars God, was none other than an ancient symbol, predating even the Egyptian civilization, which appeared as an eye and was

The "Eye of Horus," also known as "The Eye of Ra."

commonly known as the "Eye of Horus," or alternately the "Eye of Ra."

Meads also recognized this symbol from a source far more ancient than even the (possibly) 12,500 year-old Egyptian civilization—the ancient cave paintings in the Chauvet-Pont-d'Arc Cave in the Ardèche region of southern France. This cave contains some of the most (if not *the* most) ancient cave art ever found. Discovered in 1994 and dated as far back as 30,000 to 35,000 years ago, the cave art depicts various animals, scenes of hunting and the daily lives of the cave painters, and one very unique symbol—the Eye of Horus

Chauvet Cave art compared with the Egyptian "Eye of Horus."

When placed side by- ide, there can be little doubt that the two symbols are virtually identical. At the worst they are variations on the same very ancient theme. The meaning of this is profound. How exactly a sacred ancient symbol of the people of southern France could have made its way some 25,000 years later to the valley of the Nile where it likewise became an equally sacred ancient symbol is something that conventional history cannot explain. But even more profound is how this selfsame ancient symbol could have found its way perhaps hundreds of thousands of years previously onto a knobby plain on the surface of Mars.

While the connection between Horus, his symbol and the planet Mars is well established, there are some who question the true meaning of the "Eye" symbol. It has become popular, especially among New Age theorists like David Wilcock, to equate the "Eye of Horus" with a cross section of the human pineal gland, which

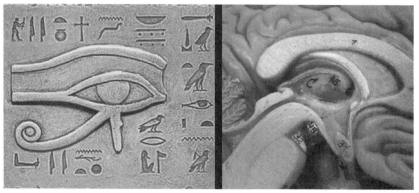

Eye of Horus hieroglyph and cross section of the human pineal gland.

they equate with telepathy and psychic abilities. While I find this possibility interesting, I am far more intrigued by the possibility that this symbol might have been sculpted on Mars and then somehow brought to Earth, which would imply an even deeper connection between Earth and the Red Planet than anyone previously suspected. The only question is if this knob on the surface of Mars is indeed artificial, or if it is simply, as NASA always likes to say, simply a trick of erosion or light and shadow.

In examining the formation more closely, it does show signs of being either artificial or at least artificially altered. The area around the "eyeball" contains a distinctly artificial looking crosshatching

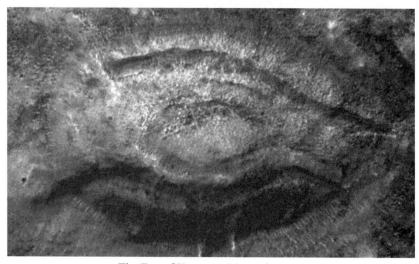

The Eye of Horus on Mars, close up.

187

pattern, which is a sure sign of an underlying superstructure. There is a prominent bar which is exposed and runs from the inner part of the eyeball to its outer rim. Parallel bars run at 90 degree angles to this central structural piece, showing a tube framed structure with smaller cross-supporting cells underneath it. More such right-angled tubular supports can be seen in the curved groove just above the eye, and just beyond the structure in the northwest quadrant there is more of this underlying rectilinear support structure. All of these identical looking structural pieces are partially buried, obviously exposed by years of wind erosion.

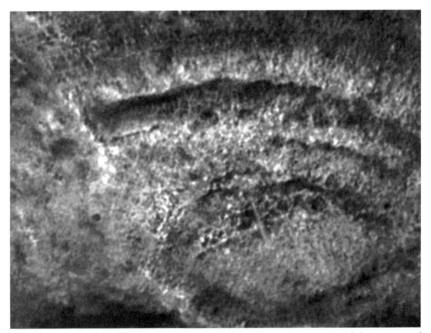

Close-up of right-angle, structural, tube-frame supports exposed by years of erosion on the "Eye of Horus."

That in and of itself would be enough to mark this formation as a candidate for artificiality, possibly even being some kind of temple or other house of worship. If that is indeed the case, it might make more sense as a universal symbol from across the eons, meant to show a connection between Earth and Mars, and possibly even ancient Egypt.

But if that's what it is—a temple of some kind—then there

would have to be some kind of obvious entrance for where the pilgrims and worshippers could enter the structure.

There is.

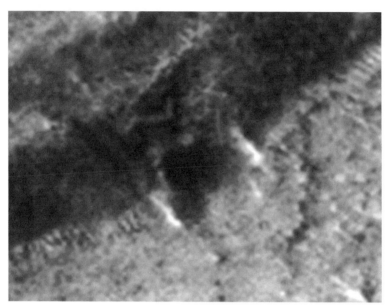

Close-up of an obvious "entrance" to the Eye of Horus.

In the lower middle of the Eye of Horus is what certainly looks like an obvious entrance to the structure. Marked by two bright supports, there is a dark arch-like entrance with numerous structural beams, girders and tubes overhead. The simple fact is that naturally eroded knobs and mesas do not have tube frame supports with crossing support structures on top of key features, nor do they have clearly marked entrances at their base for visitors to enter and examine their interiors. But artificial structures do.

So in spite of the fact that the Eye of Horus seems to be somewhat isolated when compared to other artificial structures like the Face at Cydonia, I think it is artificial. I have to say, if I could pick a landing spot for NASA's next rover, I'd put it right outside the "Eye" and send it to that dark archway straightaway to peer inside. There's no telling what we might find.

In response to discoveries like this, the critics usually offer excuses along the lines of "Mars is an alien planet. It's a really weird place." The truth is, they're right, Mars is weird. The problem is, it shouldn't be. The same rules of geological erosion apply to Mars as they do to Earth. But artifact after artifact shows that Mars is covered with objects that can't be found on Earth, except in the realm of artificial construction. And falling back on the "It's an alien planet" excuse isn't good enough when they claim that all of this can be explained geologically.

But arguing about satellite imagery endlessly can only get us so far. To really seal the deal and expose the reality of what Mars is, we need corresponding "ground truth." We got some of that from the early *Pathfinder* probe (see *Ancient Aliens on Mars*). But in the last decade, we have gotten even more from the three subsequent rovers, *Spirit, Opportunity* and *Curiosity*. It is from them that we will see the final proof of the thesis that Mars was once the home of a vast and now sadly destroyed civilization.

A civilization that NASA seems determined to investigate, while at the same time pretending never existed.

(Endnotes)

1 http://www.marsanomalyresearch.com/evidence-reports/2009/167/parro-topia.htm

2 http://spsr.utsi.edu/articles/JSE253Saunders.pdf

3 http://mysticalfiles.com/the-mars-earth-thing/

Chapter 8
The Martian Junkyard

In my previous book, *Ancient Aliens on Mars*, we discussed and examined examples of what appeared to be wrecked technology and mechanisms in the vicinity of the *Pathfinder* rover. Since *Pathfinder* in 1997, NASA has visited Mars with three more such rovers, each bigger, better and more sophisticated than the previous ones. The twin rovers *Spirit* and *Opportunity* touched down within days of each other in 2004, and *Curiosity* followed in 2012, landing not too far from Parrot City in a crater named Gale. If my core thesis is correct, then these rovers, like *Pathfinder* before them, should have found evidence of the ancient, ruined and abandoned Martian civilization that once flourished on the Red Planet. Not only did they find evidence of that, but they also found a great deal more. They found evidence of life itself, both dormant and fossilized, and some still actively growing on the surface of Mars today in the form of simple plants and lichens. In this final chapter, we will look at all of these, and then ultimately ponder what it all means for us and our own future here on Earth.

The craze about plant life on Mars started long ago with the "wave of darkening" I mentioned in *Ancient Aliens on Mars*. It took a broad step forward with the experiments of Dr. Gil Levin on the 1970s *Viking* landers, both in the form of his Labeled Release Experiment, which came back positive for microbial life today on Mars, and in the imagery of "patch rock" and other rocks at the *Viking* landing sites which showed simple, lichen-like plants waxing and waning in the Martian biosphere with the changing of the seasons. But it took another step forward when *Mars Global Surveyor* started sending back images of the surface at a resolution never seen before. It was from those images that we began to see phenomena like "Arthur's Bushes."

"Banyan Trees" on Mars, from NASA/MSSS image M0-804688, and from Wayne Newton's estate in Las Vegas.

Explained away by NASA as carbon dioxide "spouts" spewing dark material onto the snowy Martian surface, the "spiders" as NASA has dubbed them, were quickly set upon by science fiction writer Arthur C. Clarke as proof of large-scale plant life on Mars. He compared them to banyan trees here on Earth, and quickly dismissed the "spider spouts" argument, as did many others who noted the shadows and 3D depth of the trees, and how they seemed to grow together as the environment warmed in the summers on

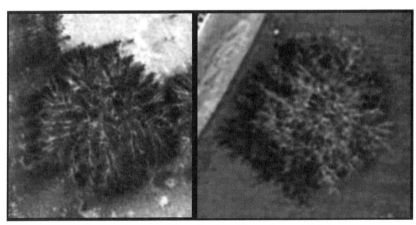

Banyan trees on Mars (left) and Las Vegas (right).

the Red Planet. Even the usual NASA lapdogs at *Wikipedia* had to admit on a page about the phenomenon that "They are unlike any terrestrial geological phenomenon."[1] That, of course, is because they *aren't* a "geological phenomenon."

A simple visual comparison between the "spiders" of Mars and the banyan trees of Earth reveals just how absurd NASAs objections are. Faced with the obvious match, they are forced to resort to the time honored "Jedi mind trick" explanation that it's all just an optical illusion. That got a bit more complicated and less believable when images started coming back from *Mars Reconnaissance Orbiter* showing tall trees with trunks that actually cast shadows on the Martian surface.

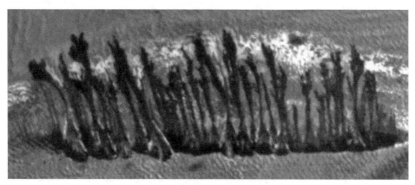
"Sand Slides" on Mars.

But NASA persisted in denying there was anything there at all:

They might look like trees on Mars, but they're not. Groups of dark brown streaks have been photographed by the Mars Reconnaissance Orbiter on melting pinkish sand dunes covered with light frost. The above image was taken near the North Pole of Mars. At that time, dark sand on the interior of Martian sand dunes became more and more visible as the spring Sun melted the lighter carbon dioxide ice. When occurring near the top of a dune, dark sand may cascade down the dune leaving dark surface streaks—streaks that might appear at first to be trees standing in front of the lighter regions, but cast no shadows. Objects about 25

193

centimeters across are resolved on this image spanning about one kilometer. Close ups of some parts of this image show billowing plumes indicating that the sand slides were occurring even when the image was being taken.[2]

So according to NASA, these were simply stains of dark sands sliding down the mountainsides, rather than waif-like, wispy trees sprouting *upwards* from the ground below. Their mind trick may have worked, were it not for the close-ups of the "trunks," which showed arched, V-shaped root forms that cast shadows on the ground beneath them...

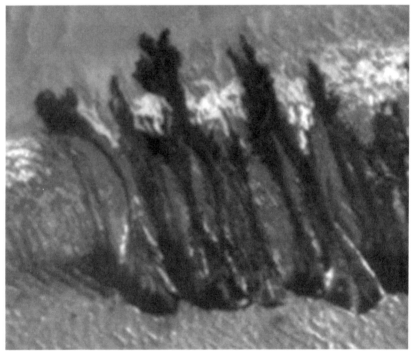

Tree trunks and bent over "palm trees" on Mars.

So these "billowing plumes" were exactly the opposite of what NASA claimed they were, and contrary to the NASA press releases they had solid, root like bases that did indeed cast shadows on the Martian surface. Clearly, this kind of thing is meant for the Space.com crowd and the "honest but stupid" folks inside NASA at the lower levels.

At any rate, this gave me a pretty good idea that NASA knew a great deal more than they were letting on about simple plant life on Mars, and this was only reinforced in looking at their choices for landing locations for the new rovers. For *Spirit*, they chose Gusev crater, a region that had shown some very strange dark streaks and splotches on its surface when it was photographed during the *Viking* era. When *Mars Express* re-imaged the area in preparation for *Spirit's* landing, these dark streaks turned out to be *bright green*, and looked distinctly like vegetation. NASA quickly messed the color up and tried to make the green appear like more of a dark grey, but the damage had been done and it seemed clear that they were landing *Spirit* in a zone they hoped might sustain extant organic life (see *Dark Mission*).

But NASA quickly got some unpleasant surprises in their surreptitious search for life on Mars. Almost as soon as the *Spirit* rover descended from its landing platform, it started sending back pictures of curious objects that were not easy to explain away as mere optical illusions.

"Slot Rock" on Mars.

The first of these was "slot rock" a beaten up, metal cased object which had a pattern of what looked like bolt holes on a very flat, sheared off face. It also had a nearly perfect rectangular opening

Mettallic gleam off the top of "slot rock."

in the front. Later images showed that it seemed to glow brightly, with a very metallic looking surface texture.

None of these characteristics—the flat face, as if it had been sheared off by a power cutter, the pattern of the "bolt holes," the metallic gleam or the nearly perfect rectangular hole—are consistent with natural geology. They are however characteristic of manufactured machinery. As the days passed, we anxiously awaited the rover's next manuver, hoping that it might drive over to slot rock and use its cameras to peer inside the hole. Instead, when *Spirit* descended the ramps and began to explore the Martian surface in detail, it *drove right past it* and directly up to what was obviously a very conventional rock. If *Spirit* did stop by and take any images of the inside of that strange rectangular opening, they were never released.

As *Spirit* began to roam the surface however, it did stop to take images with its panoramic camera of the surrounding terrain. As we had found with the *Pathfinder* images from seven years before, NASA's camera technology had taken a precipitous drop in quality

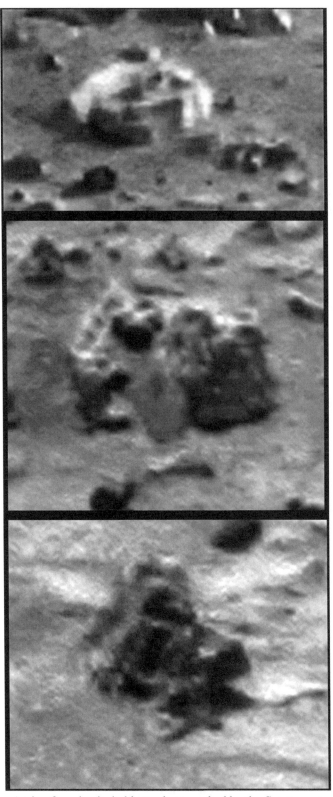

A trio of mechanical objects photographed by the *Spirit* rover.

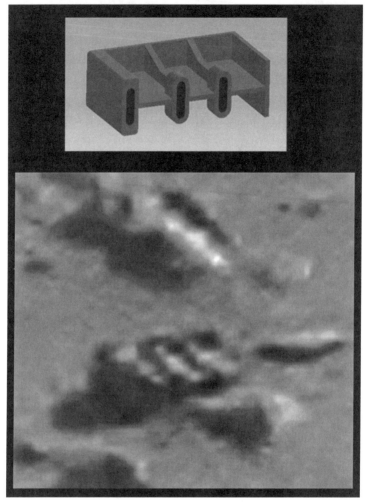

The "machined fitting" on Mars.

from the mid-1970s *Viking* missions, but the cameras still caught images of a number of very interesting surface objects.

These bits of mechanical debris were easily reconizable by their right-angled components, tubular and cylindrical extensions, squared-off access holes and sometimes partially eroded metal shell casings. Other objects diplayed flanges, base plates, regularly spaced stiffeners and regularly spaced holes or attachment fittings. All of these characteristics are telltale signs of engineering design intent.

After spending some time in the local area, *Spirit* then set off to a distant rock outcrop nicknamed "home plate."

At the same time as all this was going on, the second Martian rover, *Opportunity*, had made a successful landing in Meridiani Planum. On Sol 33, its 33rd day on Mars, the rover was commanded to roll forward to begin intensive investigation of a small section of rocky outcrop rimming the small crater that it had landed in on January 25, 2004. The outcrop, only a few inches high but which spans approximately 180° of the crater's interior, had been dubbed by the JPL Rover Team "Opportunity Ledge." The specific section that *Opportunity* was ordered to investigate was about in the middle of this outcrop, is approximately ten inches high and was named by the team "El Capitan."

One of the first discoveries made by *Opportunity* as it headed to El Capitan were rounded spherules found embedded in various rocky outcrops around Meridiani Planum. Nicknamed "blueberries" by JPL scientists because of their blue color and the way they were embedded in the rock—like blueberries in a blueberry muffin— these objects bore a striking resemblance to Pentremites, a form of plant that went extinct on Earth 250 million years ago.

Pentremite fossil (left) and Martian "blueberry" (right).

NASA eventually declared that the spherules were proof that liquid water once flowed on Mars, and that they had formed in place. After that, not much was said about them and *Opportunity* continued on its way to El Capitan. Preparatory to actually drilling into El Capitan, *Opportunity* was commanded to take a series of close-up images of the untouched surface of the rock with the

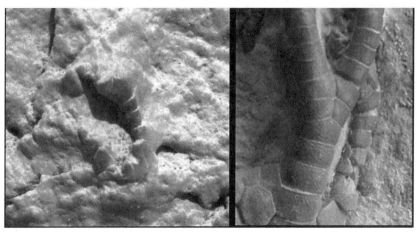

Segmented, crinoid-like Martian fossil (left) compared to actual Crinoid fossil (right). NASA image 1M131201699EFF0500P29933M2M1.jpg.

grayscale microscopic CCD camera attached to its arm. One of those images revealed an amazing sight: an apparent Martian fossil.

Found right next to more of the embedded "blueberries," a close-up enlargement revealed an apparently snapped-off pair of limbs, at least five visible cylindrical segments, and a hint of other fossil-like features buried in the surrounding rock itself—all classic hallmarks of a formerly living organism. One collector of marine fossils, James Calhoun, agreed immediately that it was a fossilized life-form:

> I have been a collector of marine fossils for thirty-four years, with years of field experience. When I saw the 'Fossil' pic, it was clear to me that it met a number of the basic criteria of fossilization… I have heard a varied number of explanations as to what type of fossil this could be, everything from a segmented worm (annelid) to a shrimp (crustacean). I would like you guys to consider that based on the symmetry of the object, that it could be in fact an early Crinoid, a filter feeding marine plant-like animal, a type with a calcium carbonate exoskeleton. I have included a couple of pictures for symmetry and scale reference. Notice the triangular symmetry in the 'branch areas,' not to mention the segments, and also that the scale is inline with the 'size of the blueberries.'

The images Calhoun included with his e-mail were a perfect match for the Martian fossil.

A Crinoid (sometimes called a "sea lily," because of its superficial similarity to a spreading flower) is, as Calhoun described, "a filter feeding marine plant-like animal." Crinoids first appeared in Earth's primeval seas over 500 million years ago, in the so-called "Cambrian Explosion," when millions of new life forms suddenly appeared on Earth seemingly out of nowhere. Climbing to dominance over the next 150 million years, the Crinoids then began to recede in the terrestrial fossil record.

Crinoids lived in ocean water—ranging from a few feet deep to several miles—anchoring their stems on the ocean floor and feeding on whatever nutrients drifted by. If you look at a combined map of where JPL landed its two rovers and the *Odyssey* Gamma Ray Spectrometer orbital determination of water abundance in the upper one meter of Martian soil, a glance will suffice to show the rovers are indeed exploring none other than the shallows of two proposed equatorial Martian tidal oceans (see the Mars Tidal Model in *Ancient Aliens on Mars*). It takes almost no imagination to picture this site several million years ago as a quiet tidal pool, filled with gently waving creatures of the sea, until one day something extraordinary and cataclysmic happened, and this shallow pool and all of Mars was forever changed. It was now clear that NASA had picked the two rover landing sites because they presented an opportunity (pun intended) to search for signs of life were water once flowed.

So, upon making this extraordinary discovery, nothing less than proof of life on Mars, at least in the distant past, what did NASA and the *Opportunity* rover team promptly do? Did they call a press conference and hail their discovery to the world? Did they head to the White House to brief the president on the most momentous scientific discovery in all of human history? No. They immediately took *Opportunity's* drill (technically called the "Rock Abrasion Tool," or RAT), and promptly *ground it into powder*. Instead of moving the grinder a couple inches to the left or right, they simply bored down on the fossil, totally obliterating it.

And learning nothing.

There is no question that the rover team saw the crinoid fossil. Not only did they use it as the base target for the drill, they gave it a target name: "Guadalupe." Our Lady of Guadalupe is a Roman Catholic religious icon, functionally equivalent to the Egyptian Isis, the goddess of life. Yet after discovering a fossil that could confirm the existence of complex life on Mars sometime in the past, and even naming it after a "goddess of life," they destroyed it. Since *Opportunity's* instrument suite was not designed to look for signs of life (it was strictly set up to be a roving geologist), Guadalupe's destruction served no real scientific purpose. It had long ago taken on the properties of its surrounding rock. The only testament to the fact that it once had lived was that unmistakable segmented shape in the rocks—and NASA destroyed it.

Since then, a former NASA employee named Richard Hoover has been making the rounds at various UFO expos and conferences. He frequently brings up the Crinoid fossil, as if he is the one who found it, but never mentions that it was first noted and presented by Richard C. Hoagland. I wish to make it clear in these pages that no matter what you may hear at these conferences or from Mr. Hoover himself, it is well established that Hoagland was the one who found it and first wrote about it.

"Dinosaur skulls" from *Spirit* PanCam image 2P127793693EFF0327P2371R1M1.jpg.

Spirit meanwhile, was on the move and almost as soon as she began her long trek to home plate she began to image a variety of objects that stood out from the normal background of plain rocks and geology. On Sol 16 (*Spirit's* 16th "day" on Mars), the Panoramic camera photographed a nearby field of objects that contained at least two distinct fossilized animal skulls. There were also numerous other "rocks" in the image that looked much more like bone fragments than geology, but the two lizard/birdlike skulls were the most obvious. First reported on by Joseph P. Skipper of Mars Anomaly Research on September 5[t], 2004,[3] the two skulls were near exact matches for dinosaur skulls found on Earth.

One, in fact, was an excellent match for a Bagaceratops, a three foot long herbivore which lived during the late Cretaceous period about 80 million years ago. In fact, the skull found in Gusev crater on Sol 16 is a near exact match in structure, length and volume to juvenile Bagaceratops skull samples. The other more birdlike skull

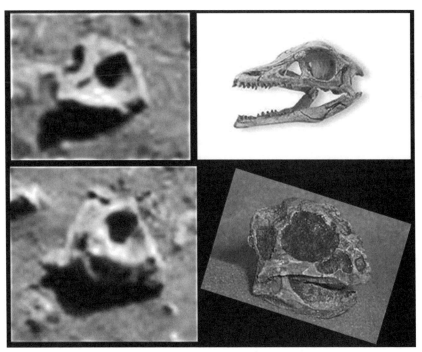

Dinosaur skulls found on Mars, and terrestrial comparisons.

Humanoid skull amidst rock field from Sol 513 NASA PanCam image 2P171912249EFFAAL4P2425L7M1.jpg.

is similar to—but not an exact match for—an early bird species named Archaeopteryx, which lived about 150 million years ago, during the early Tithonian stage of the Jurassic period. Isolated, they might easily have been dismissed as merely weird rocks on an alien planet, as "Dr. Phil" and the debunkers like to claim. But taken together and in such close proximity, they are much harder to dismiss, especially given what *Spirit,* and later *Curiosity,* were about to encounter.

    Spirit continued its long journey to home plate and was nearly 75 percent of the way there when it encountered something on a PanCam image that shocked the entire independent Mars research community—a humanoid skull.

    First noticed on *Spirit* Sol 513 images by Eduardo Lucena in Brazil and then later by Joseph P. Skipper of Mars Anomaly Research,[4] the Martian skull remains one of the hardest-to-refute biological finds by any of the rovers. Standing out against an otherwise "normal" looking field of rocks, the skull looks like some

Humanoid skull from Sol 513 image 2P171912249EFFAAL4P2425L7M1.jpg

kind of cross between a human skull and a storm trooper helmet from *Star Wars*.

Under normal circumstances, it might be dismissed as a geologic anomaly, but upon closer examination it has all the earmarks of biology rather than geology. The skull is rounded and well formed, very similar to a human skull. There are two equally spaced eye sockets, a bone bridge between the eyes and a nasal protrusion with what look to be holes for some kind of nostrils. The mouth area is where it really begins to look different from a typical Homo sapiens however. The Martian skull has a gaping mouth and enormous jowl-like appendages attached to the lower skull. Whoever this guy was, it's a safe bet that, to quote Arnold Schwarzenegger in *Predator*, he was "one ugly motherfucker."

There are also other details which lead me to believe this is not a rock, but the skull of some long dead alien creature. Right above the right eyesocket are a series of dark indentations along the cranium, separated by raised, white, bony structures. Like the indentations in our own human skulls, these appear to be the attachment points for some very heavy muscles and tendons, which

205

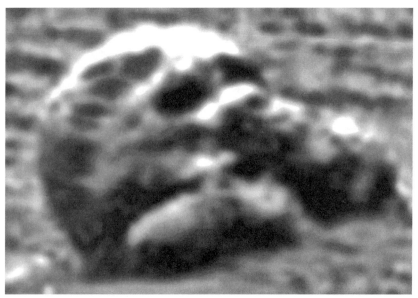

The skull with dark cranial indentations over-emphasized for clarity.

would make biological sense given the great weight the jowl-like extensions would place on the head and neck. It's hard to believe such useful evolutionary features could have made their way onto a mere "rock" just by pure coincidence.

All in all, I feel strongly that based on this one picture alone, there is sufficient evidence to conclude that the humanoid skull is exactly that—though it is also safe to say I wouldn't want to meet the owner of it in some proverbial dark alley someday. Sadly, *Spirit* didn't linger and take more pictures of the humanoid skull but instead moved on to home plate, where things, as it turned out, got even more interesting...

Home plate is large rocky outcrop so named because of its similarity in shape to a baseball field's home plate and because of its white color. Standing out against the background of fairly innocuous terrain, it's easy to see why NASA would be interested in it, even from a purely geological perspective. It was even more obvious after *Spirit* arrived.

From early on when the mission plans were announced, independent researcher Richard C. Hoagland (my co-author on *Dark Mission*) had argued that home plate was not a natural

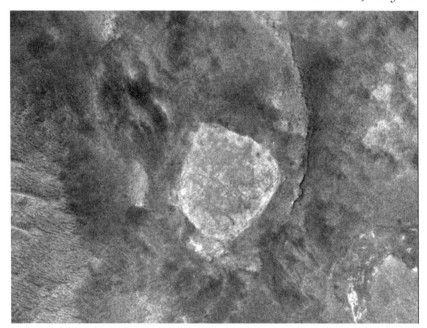

MRO image of "Home Plate"

geological formation, but rather the eroded foundation of some much larger artificial structure. Citing the fact that there were few (if any) terrestrial comps for such an unusual "rock outcrop," Hoagland predicted that home plate would become a treasure trove for a Martian salvage reconnaissance mission. If there was junk to be found within the range of one of the rovers on Mars, he reasoned, home plate would be the place to be rooting around.

How right he was.

After arriving at home plate, *Spirit* made its way carefully around the perimeter before settling in for the winter. After the Martian winter had passed, the rover emerged from its state of hibernation and made its way down onto the surface of home plate itself. As it did so, the rover was stopped and took a high resolution panorama of the area, image number PIA10214. Almost immediately after its release, a huge controversy erupted.

In the left hand corner of the image was a small shape which some felt resembled a human figure with its arm outstretched. They believed it resembled a sculpture of a mermaid or other humanoid form.

The "Martian Figurine" from PIA10214.

The reaction was immediate. CNN, Fox News and the other news networks immediately reported on the "Mermaid on Mars," mocking the idea that it could be anything but a "funny looking rock." In this they were assisted by the usual suspects, people like James Oberg and Phil "Dr. Phil" Plait, who viciously attacked anyone who even felt the object should be closer investigated. Plait, always quick to dissuade the scientific investigation of anything he doesn't believe in by any means necessary, including, as we have learned in this volume, outright deception, was especially scathing in his attacks.

In a Space.com article authored by a *Skeptical Inquirer* reporter named Benjamin Radford, Plait even went so far as to accuse researchers of thinking the figurine was a representation of Bigfoot:

According to astronomer Phil Plait of the Bad Astronomy Web site, if the image really is of a man on Mars, he's awfully small: "Talk about a tempest in a teacup!" Plait said. "The rock on Mars is actually just a few inches high and a few yards from the camera.

*A few million years of Martian winds sculpted it into an odd shape,
which happens to look like, well, a Bigfoot! It's just our natural
tendency to see familiar shapes in random objects.*[5]

Of course, no one in the independent research community
claimed it was a statue of Bigfoot, but that didn't stop Plait from
making the claim. Also lost in his commentary was the fact that if
"a few million years of Martian winds" had actually sculpted the
figurine, there should have been examples of other such smooth and
unusual shapes nearby. There were not.

Finally, Plait fell back on the idea that the object was simply
too small to be a real sculpture, since it was only "a few inches high
and a few yards from the camera." In reality, this argument is idiotic
on many levels.

First of all, scale is not a factor in such an evaluation since
scale is not a factor in sculpture. As an example, there are many
Egyptian sculptures as well as sculptures from other ancient cultures
that scale out in exactly the same manner. In terms of size, the
Martian figurine compares very favorably with ancient sculptures
of the Pharaoh Khufu, for example.

Second, Plait once again falls back on the discredited
"pareidolia" argument, with the author of the article claiming
that pareidolia "is well known in psychology," which we have
already established is an utter falsehood. So, if the figurine isn't

Comparison of the Martian figurine with an Egyptian sculpture of
the Pharaoh Khufu. Objects are to scale.

The Little Mermaid statue in Copenhagen harbor compared to the Martian figurine.

Bigfoot, and it isn't too small to be a sculpture or a figment of our imaginations, then what exactly is it?

For starters, there is no reason it could not be a sculpture. Not because of its size, nor because of its aesthetics. It looks like a humanoid figure and it even has what looks like a base or platform upon which it rests. It bears a striking resemblance, at least in my opinion, to the Little Mermaid statue in Copenhagen harbor. Both seem to be kneeling, with one arm crossed over and the head looking off to one side. In their article Radford and Plait also tried to claim that it made no sense for the alleged sculpture to be human in appearance, since Mars was an alien planet. But it is also true that there is no reason to assume that life on Mars in the past was not related genetically to life on Earth in the present. The Crinoid fossil is an example of the possibility that both worlds may share a common genetic ancestry. So in the end, the simple fact is that there is no reason to arbitrarily dismiss the idea that it is a sculpture.

But if the figurine was completely isolated, if there was nothing else at all unusual around it near home plate, then the critics might have a point. However, closer scrutiny of PIA10214 shows that they don't. All around the panorama, we can see scattered signs of technology in ruins. It is actually a fairly easy task to differentiate these artificial objects from the normal rocks of the Martian landscape. Artificial mechanical constructs of the modern technological world share many things in common. They tend to have geometric shapes, like squares and rectangles and cylinders; they tend to be encased in thin sheet metal casings; they frequently have square or rounded fittings and tabs as attachment points to other mechanisms; and they usually have vents or holes for cooling of the internal mechanisms, whatever they may be. Rocks have none of these characteristics.

An "air conditioning unit" on Mars.

There are so many of these of types of mechanisms on PIA10214 that they almost outnumber the simple rocks. The "air conditioning unit" shown above is a classic example. It has a rectangular shape, a thin casing, a hint of vertical striations for

Bent sheet metal shell on top of a squared object.

211

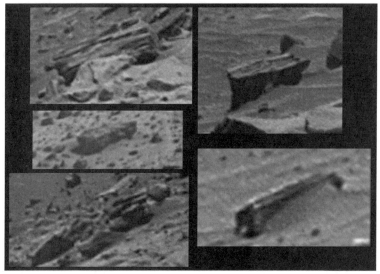

Gallery of partially buried technological objects from *Spirit* panorama PIA10214.

venting or heat sinking, and two fixed square attachment points on top. It has obviously been tossed about and beaten up, and is partially buried in the sands. But it isn't a rock.

Another example is a squared, partially buried object near the Figurine. Its thin sheet metal shell has been pried upward by some force, and you can see the shadow that this thin metal skin casts on the main structure underneath. Rocks, which are solid objects, simply don't erode in thin, metal-like sheets this way.

There are many other such box-like, rectangular objects

The "Spoke."

212

scattered all around the panorama. They have a metallic gleam (more obviously visible in the full color panorama), casing shells and slots and vents, the obvious signatures of technology. But there are also other strange objects, especially one I call the "spoke," which simply cannot be natural.

The first thing that caught my eye about the spoke was that it had a central circular hole (that's something else plain old rocks rarely have unless they've been drilled on) and a series of equally spaced and equal length fan blade type extensions emanating from the central circle. I also noticed that as NASA has done many times before, they deliberately wallpapered over the center of the object, trying to break up the lines and make its artificiality less obvious. It looked almost like a big spider lying on the Martian surface, ready to pounce on *Spirit* if it got too close.

After this panorama was taken, *Spirit* quickly moved on from the Figurine without taking any close-up images of it. As *Spirit* moved around home plate, it closely studied some even stranger objects that it encountered.

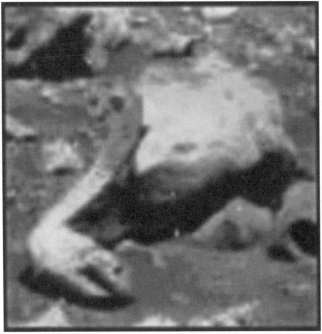

The "Wrench" from *Spirit* PanCam image taken on Sol 527.

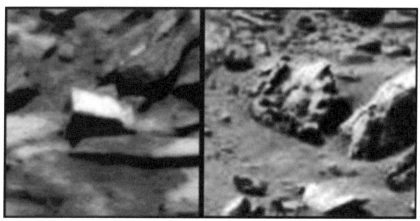

The "Cowbell" and the "Gearbox," from Sol 288 on Mars.

One of these, found by Rolf Varga on a *Spirit* Panoramic camera image taken on Sol 527, is unquestionably a sheet metal implement. Nicknamed the "Wrench" because it has a head shaped like pliers or clippers of some sort, the object is twisted and bent

Eroded sheet metal cased squared objects on Mars. Note the shiny, metallic hinged plate in the upper right, and the "projector" to the left.

214

over a prominent rock. The tail end is shaped like an arrowhead. There is even a hole in one of the leaves of the "clipper."

All these characteristics identify this object as a formed or cast sheet metal part or assembly, rather than geology. Rocks simply do not form in thin, metallic strips with functional, geometric ends. Nor do they form gears and "cowbells," like other objects *Spirit* encountered on its journey to (and over) home plate.

These were odd enough, but other undeniable artifacts were also discovered along the way. These took all shapes and sizes, from hollow, rectangular formed "cowbell" like objects to complex gear shaped machinery, which would have had to be machined and assembled. But all of these objects were more mechanical in nature than advanced technologically. Where were all the wires and electronics and circuit boards that this advanced culture would have had to produce?

We had only to wait.

Very quickly *Spirit* made its way to a hilltop area designated

Top down view of sheet metal cased objects with eroded circuit boards inside.

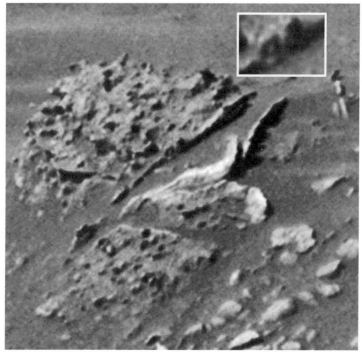

Eroded metal cased object found on home plate. Note the dual hinge fittings (inset).

"Site 132," atop home plate, and there it found even stranger artifacts. The area was littered with eroded, square, box-like shapes, partially buried in the sand. The thin eroded casings of their sheet metal boxes jutted up from the surface, and inside were a series of knobby, eroded layers of material. Nearby one of the "boxes" was a shiny metal object with what looked like a hinge fitting on it, and there was also an eroded, buried square object with a rounded "lens" that looked like a modern day computer projector.

In looking at the erosion patterns of the "guts" of these cased objects, they looked remarkably like computer component boards. They had connectors, transistors, and all manner of knobby extensions which compare very favorably to the components of modern day computers like graphics cards, TV adaptors and wireless modems. Other observers compared them to the internal guts of a heat exchanger. One of the objects even had two side-by-side hinge fittings, complete with mounting holes.

Simply put, naturally eroded objects do not look like this. However, this is exactly what discarded computer equipment and mechanical debris would look like if you tossed them around and let them sit in the Martian biosphere for a few thousand years. They would fill with dirt and silt, freeze and rust, and eventually the metal shells would erode away, leaving only the sides and the thickest parts sticking up out of the sand. Eroded circuit board components would also be confined only to those areas within the perimeter of their metal cased shells, and that is also exactly what we see in these *Spirit* images. If these were just oddly eroding rocks, then the weird knobby extrusions would be scattered all over the area in question. Instead, they only appear inside the metal casing shells.

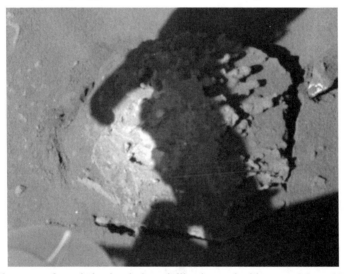

Close-up of eroded, circuit board like layered object on Mars. Note multiple cylindrical "connectors."

Such layered, fossilized technological ruins seemed to be of great interest to NASA. They took many pictures of them and in each case, the objects looked more and more like computer components or mechanical devices scattered around the home plate area. But if my core thesis is correct, that Mars is one big junkyard of technological secrets left behind by a vast Ancient Alien civilization, then there would have to be one final test. If *Pathfinder*, *Spirit* and *Opportunity* had found artifacts and ruins, then so too would the upcoming *Curiosity* lander in Gale crater. We had only to wait for the pictures. Happily, we were not at all disappointed in what *Curiosity* uncovered, and still continues to uncover to this day...

Curiosity

When the *Curiosity* rover (officially part of the Mars Science Laboratory mission) successfully landed in Gale crater on August 6, 2012, just south of the Martian equator, many of us in the independent research community had high hopes for the mission. The area was in the transitional region between the heavily cratered southern hemisphere and the much smoother northern hemisphere of Mars. This meant that it might be an area where remnants of biological life might be found and it also might be an excellent location for finding the remains of a once vast Martian civilization. Within a few hours, the rover had started to send back enough images with the powerful Navcam cameras of the area around it for NASA to put together a panorama of the landing site. That panorama, PIA16011, showed a full 360 degree view of the area around *Curiosity*. Almost due north of the landing site, along the rim of the crater, was the first sign that our expectations were not unfounded. There—plain as day—was a massive Egyptian style pyramid standing out against the background.

This "Bent" pyramid bore an eerie resemblance to the Bent

"Bent pyramid" along the northern rim of Gale crater from NASA panorama PIA16011.

"Bent" Pyramids at Dahshur, Egypt (left) and on Mars in Gale crater (right).

Pyramid in Dahshur, Egypt. Both objects had a base that started out at one angle, topped by a second upper section that narrowed to a point at a sharper angle. There was also what might have been a smaller satellite pyramid next to the Gale pyramid, exactly the same as at Dahshur. Detailed enhancements showed what may have been dark openings both at the base and on the second level of the Gale pyramid. The second level openings were distinctly rectangular and separated by brighter vertical pillars.

The comparison between the two structures on different worlds could not have been more exact, nor more compelling. As *Curiosity* made its way around Gale crater as it worked toward the

Geometric object with rectangular slot cut in it discovered by Robert Morningstar.

The Wheel, a hubcap type object with a cylindrical valve stem sticking out at 90 degrees.

central peak, we hoped to get better images of the Bent Pyramid. But curiously (pun intended), each time it had a chance to image the object, the results came back far more blurry and less clear, even though *Curiosity* was actually much closer to the pyramid.

But as the traverse continued, independent researchers found more and more objects which simply did not belong on any kind of natural Martian landscape. There were rectangular objects with multiple facets at right angles to one other, some of which even had rectangular slots cut out of them.

Sheet metal casing with regularly spaced holes punched into it.

Other findings were visually different but equally manufactured in their genesis. One wheel like object discovered by Richard C. Hoagland on a Sol 164 PanCam image looks like a battered and discarded wheel from a large truck, complete with a hollow, cylindrical "valve stem" jutting out from the exact center at a 90° angle.

Eroded sheet metal is also in abundance in Gale crater, most of it partially buried and squared off, like the walls of a metal box that has rusted away. Sheet metal pieces simply stick out of the sands of Mars, part of much larger objects that have simply eroded away.

Other objects are shiny and more clearly mechanical in nature, and come complete with design features like heat sinks, flywheels and tubular extensions...

More eroded sheet metal sticking out of the Martian sands.

One of my favorites is the "Turbocharger," a mechanical device with machined sides, heat sink fins and what may be a flywheel type attachment on one side.

Images taken on more recent Sols haven't failed to disappoint either. On a recent panorama designated PIA16700, numerous artifacts can be found, including a suitcase sized object I have

Various shiny, metallic cased objects scattered around Gale crater.

nicknamed the Tank, and a cylindrical body that shows evidence of machining all over its surface. It looks a bit like the central shell of a small jet engine.

And *Curiosity* has not disappointed in the dinosaur fossils arena either. Like *Spirit* and *Opportunity* before her, *Curiosity* has found evidence of simple plant life and complex animal life that once roamed Mars.

The "Turbocharger," a metallic object with heat sink fins on its surface.

The Tank and the Cylinder, two more clearly mechanical devices found by *Curiosity* in Gale crater (PIA16700).

Researcher Robert Morningstar has found an image that includes block-like piles of rubble, similar to the H-block ruins at Puma Punku, Bolivia. Among them, he found other blocks that had a fuzzy, greenish plant life growing on them. Just like rocks from as far back as the *Viking* landing sites in the 1970s, these clearly show signs of not only artificial origin, but also active plant and moss growth as well.

Other images, whose discoverers I have not yet determined, show what undeniably appear to be the fossils of large, serpent

Fuzzy, moss covered rock on Mars.

like or dinosaur like creatures on Mars.

These beasts, whenever they lived, certainly bear something in common with our own dinosaur fossils. Like them they are long and tube-like, and have a nearly identical skeletal structure. That implies that they are the result of some kind of biological evolution, different from but similar to that on Earth.

The nearly identical match in spinal structure strongly implies a nearly identical match in biological function. But it is impossible to know whether these Martian "dinosaur fossils" are more modern, dating from the time of the Martian catastrophe that devastated the planet, or whether they are much older than that, but were just exposed by the horrific events which wiped out Mars' civilization. We may never know unless we send men to Mars and bring some of

Comparison of Martian spine fossils (above) and dinosaur fossils on Earth (below).

Spinal bones from Mars (above) and Earth (below).

these fossils back for study.

The truth is, I could go on and on like this for page after page. There are so many bizarre, obviously technological, biological and completely unnatural objects on the various *Curiosity* images that it would take literally years to properly catalog all of them. I think at this point, we can simply state that Mars is nothing if not a giant junkyard of ruined technology, literally just waiting for us to send men there and retrieve some of these objects and declare once and for all that we are not alone, and our history is not what we have been taught in the books.

Can any rational person really discount that we might be looking at exactly that? Once you accept the notion that Mars was inhabited by an advanced civilization—a fundamental premise I think has been proven in these two books—then nothing is off the table. Not dinosaur fossils, Crinoids, humanoid skulls, Martian junkyards or massive, monumental Faces on Mars. It is all more than possible. In fact, when taken in its entirety, it paints a compelling if not airtight case for the underlying thesis.

We are left then, with only one final question: What does it all mean?

(Endnotes)

1 http://en.wikipedia.org/wiki/Geyser_(Mars)

2 http://commons.wikimedia.org/wiki/File:Dark_Sand_Cascades_on_Mars.jpg

3 http://www.marsanomalyresearch.com/evidence-reports/2004/073/animal-skulls.htm

4 http://www.marsanomalyresearch.com/evidence-reports/2006/102/mars-humanoid-skull.htm

5 http://www.space.com/4876-female-figure-mars-rock.htm

Epilogue

We are now, after two books, left with some very intriguing questions. Yet, we also have some answers. We know that, just as the Hopi said in their legends, Mars and perhaps even the entire solar system were once inhabited by a far-reaching, technologically advanced civilization. We know that, just as the Egyptians always believed, there is some connection between that antediluvian civilization and ours. We know that there were once at least two more planets in our solar system, one that Mars orbited, named Maldek, and one between Mars and Jupiter, known as Phaeton. We can be very sure that they were probably what astronomers now call "Super Earths." These massive yet unstable planets were not unlike our own terrestrial planet Earth. One or both of these lost worlds may have even been capable of supporting life. But sometime in the past—perhaps very long ago, in the millions of years—these planets were destroyed in some sort of cataclysms that left only the asteroid belt and Mars' pockmarked surface as their signatures. These disasters wiped out most of the Martian civilization, with the few survivors probably getting by for a time in underground cities they constructed after the first disaster; something like the Krell race in the classic sci-fi film *Forbidden Planet*. But nothing could save them from the second disaster, and at that point they must have had to move on to the only place left in the solar system capable of supporting advanced life—Earth.

Once here, they probably found that the best way to adapt to this new world was to intermingle their genetic code with that of the local inhabitants, and thereby create a new species we now know as Man. Certainly, Zecharia Sitchin's stories of an extraterrestrial race of giants living among us has a certain kind of relevance to it; a certain feeling of comfort and likelihood, almost as if it is some kind of shared, racial memory. His Annunaki may very well have been the Martians, and perhaps the ancient laboratories on the dying Mars made that world the "Planet of (the) Crossing" the ancient Sumerians wrote about, and called Nibiru.

Man himself has lived a tumultuous life here on Earth, probably rising and falling many times as the Hopi have told us, but he has managed to crawl back up the evolutionary ladder to the point that we can now visit Mars once again, and glimpse our once— and perhaps our future—home.

Those are my thoughts, at least.

But I want to leave you with another thought, an idea that may well lead you to conclude that this speculative history may be more than that—it may be an axiomatic truth. And there may be many people among us who have known this truth for a very

Jack Kirby

long time, and have worked to preserve it until the time was right to return to us that which is our true heritage. Our true place in the heavens.

Mars.

I come to this idea not on my own, but courtesy of a well-known and well respected master of popular culture who has created some of the most memorable characters in pop literature in the last half-century. His name is Jack Kirby.

Jack Kirby was a comic book writer and artist who worked for both DC Comics and Marvel comics over the years, and created or co-created characters and comic series like the Fantastic Four, Captain America, the Hulk and the X-Men. It is safe to say that without Kirby, it is unlikely we would have much in the way of

"The Face on Mars" from the Race to the Moon comic book series.

summer movies to spend our cash on these days.

But in 1958, 18 years before the NASA *Viking* missions to Mars, Kirby was a struggling young artist with a wife and family, and was doing as much freelance work as he could. He was commissioned by a company named Harvey Comics to write and draw a series of comics for them entitled "Race for the Moon." The Race for the Moon comics each contained three to four stories about space travel and what Man might encounter in his explorations of the solar system. Issue #2 had one particular story which has haunted me for years. It was entitled simply, "The Face on Mars."

In the story, the United States' first Mars exploration team, "Mars Expedition 1," comes across a compelling artifact—a giant, human looking sculpture of a Face on Mars. The intrepid team of astronauts proceeds to climb the structure, much as I'd love to do some day. Eventually, they reach the area around the right eye, and one of the explorers, named Ben Fisher, decides to enter it through a hole in the eyeball.

Panels from "The Face on Mars" (1958).

After falling a distance that should have been enough to kill him, Fisher finds himself in a lush, garden like paradise. He soon discovers that this paradise is inhabited by a race of peaceful and highly advanced "magnificent giants." As he moves through their world, no one seems to take note of him and Fisher soon realizes he is seeing some sort of projection—a holographic replay of some distant past events.

This realization and his happy mood are soon shattered when the peaceful race of Annunaki-like giants are suddenly attacked by a fleet of spaceships. He watches as the giants' civilization is

The attack of the machines.

devastated by the attack and discovers that the attackers are a race of intelligent, though malevolent, machines.

Devastated by the surprise attack, the magnificent giants are forced to move underground, where they plan and launch a counter attack that will destroy their enemy but leave what's left of their world uninhabitable. Fisher observes the plan being laid out and sees the leader of the giants—who plans to sacrifice himself in the attack—pointing to the home world of the enemy.

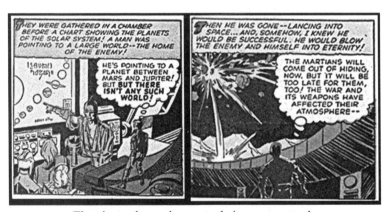

The giants plan and execute their counter attack.

The planet he points to is Phaeton, the Super Earth that once existed between Mars and Jupiter. Once the plan is carried out, the debris strikes Mars and renders it uninhabitable. Fisher then begins to choke on the thin Martian atmosphere, and is rescued by his colleagues and they continue on to Jupiter, passing through the asteroid belt along the way. Fisher and one of the other astronauts, Brennan, have an exchange about the belt, with Brennan saying that there's a theory the asteroids are the debris of a planet. Fisher comments, "It must have been the home of MONSTERS—to have suffered such a fate!"

Final plate of the Face on Mars comic.

So here, in a comic book written nearly two decades before anyone had heard of Sitchin or giants or the Face on Mars, is the whole history laid out. It makes clear that the destruction of Mars' biosphere—one of the most enduring mysteries of science—was caused by a war between a race of peaceful giants and machines. Machines which quite probably were their own creations, who eventually got too smart and turned on their masters with catastrophic results, for themselves, for the giants, for Mars, and quite possibly for Earth as well.

It was almost as if somebody knew something.

Maybe they did. Maybe they told Kirby the whole story, and had him lay it all out as some kind of cultural conditioning for the day the whole story could be told. Or maybe he just channeled it in some way, tapping into a racial memory that is embedded in the DNA we inherited from that race of giants. Maybe that is what drives us to go back to Mars and try to solve the age-old mystery that its very presence in our night sky represents, and the question it invariably forces us to ask: what happened to Mars?

Officially, we may never know. Officially we may never be told. But the really cool thing about Truth is that sooner or later, no matter what the entrenched interests say or the established orthodoxies demand, it always comes out. For us, the key to that truth lies just where Kirby said it was in his final panel of the book in 1958: "I didn't explain it all to Brennan! He and all Mankind would learn it, someday…

…From the Face on Mars."

ANCIENT ALIENS ON THE MOON
By Mike Bara
What did NASA find in their explorations of the solar system that they may have kept from the general public? How ancient really are these ruins on the Moon? Using official NASA and Russian photos of the Moon, Bara looks at vast cityscapes and domes in the Sinus Medii region as well as glass domes in the Crisium region. Bara also takes a detailed look at the mission of Apollo 17 and the case that this was a salvage mission, primarily concerned with investigating an opening into a massive hexagonal ruin near the landing site. Chapters include: The History of Lunar Anomalies; The Early 20th Century; Sinus Medii; To the Moon Alice!; Mare Crisium; Yes, Virginia, We Really Went to the Moon; Apollo 17; more. Tons of photos of the Moon examined for possible structures and other anomalies.
248 Pages. 6x9 Paperback. Illustrated.. $19.95. Code: AAOM

ANCIENT ALIENS ON MARS
By Mike Bara
Bara brings us this lavishly illustrated volume on alien structures on Mars. Was there once a vast, technologically advanced civilization on Mars, and did it leave evidence of its existence behind for humans to find eons later? Did these advanced extraterrestrial visitors vanish in a solar system wide cataclysm of their own making, only to make their way to Earth and start anew? Was Mars once as lush and green as the Earth, and teeming with life? Chapters include: War of the Worlds; The Mars Tidal Model; The Death of Mars; Cydonia and the Face on Mars; The Monuments of Mars; The Search for Life on Mars; The True Colors of Mars and The Pathfinder Sphinx; more. Color section.
252 Pages. 6x9 Paperback. Illustrated. $19.95. Code: AMAR

ANCIENT TECHNOLOGY IN PERU & BOLIVIA
By David Hatcher Childress
Childress speculates on the existence of a sunken city in Lake Titicaca and reveals new evidence that the Sumerians may have arrived in South America 4,000 years ago. He demonstrates that the use of "keystone cuts" with metal clamps poured into them to secure megalithic construction was an advanced technology used all over the world, from the Andes to Egypt, Greece and Southeast Asia. He maintains that only power tools could have made the intricate articulation and drill holes found in extremely hard granite and basalt blocks in Bolivia and Peru, and that the megalith builders had to have had advanced methods for moving and stacking gigantic blocks of stone, some weighing over 100 tons.
340 Pages. 6x9 Paperback. Illustrated.. $19.95 Code: ATP

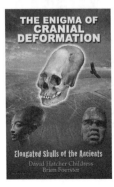

THE ENIGMA OF CRANIAL DEFORMATION
Elongated Skulls of the Ancients
By David Hatcher Childress and Brien Foerster
In a book filled with over a hundred astonishing photos and a color photo section, Childress and Foerster take us to Peru, Bolivia, Egypt, Malta, China, Mexico and other places in search of strange elongated skulls and other cranial deformation. The puzzle of why diverse ancient people—even on remote Pacific Islands—would use head-binding to create elongated heads is mystifying. Where did they even get this idea? Did some people naturally look this way—with long narrow heads? Were they some alien race? Were they an elite race that roamed the entire planet? Why do anthropologists rarely talk about cranial deformation and know so little about it?
250 Pages. 6x9 Paperback. Illustrated. $19.95. Code: ECD

VIMANA:
Flying Machines of the Ancients
by David Hatcher Childress
According to early Sanskrit texts the ancients had several types of airships called vimanas. Like aircraft of today, vimanas were used to fly through the air from city to city; to conduct aerial surveys of uncharted lands; and as delivery vehicles for awesome weapons. David Hatcher Childress, popular *Lost Cities* author and star of the History Channel's long-running show Ancient Aliens, takes us on an astounding investigation into tales of ancient flying machines. In his new book, packed with photos and diagrams, he consults ancient texts and modern stories and presents astonishing evidence that aircraft, similar to the ones we use today, were used thousands of years ago in India, Sumeria, China and other countries. Includes a 24-page color section.

408 Pages. 6x9 Paperback. Illustrated. $22.95. Code: VMA

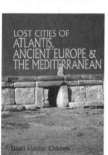

LOST CITIES OF ATLANTIS, ANCIENT EUROPE & THE MEDITERRANEAN
by David Hatcher Childress
Childress takes the reader in search of sunken cities in the Mediterranean; across the Atlas Mountains in search of Atlantean ruins; to remote islands in search of megalithic ruins; to meet living legends and secret societies. From Ireland to Turkey, Morocco to Eastern Europe, and around the remote islands of the Mediterranean and Atlantic, Childress takes the reader on an astonishing quest for mankind's past. Ancient technology, cataclysms, megalithic construction, lost civilizations and devastating wars of the past are all explored in this book.

524 PAGES. 6X9 PAPERBACK. ILLUSTRATED. $16.95. CODE: MED

LOST CITIES OF CHINA, CENTRAL ASIA & INDIA
by David Hatcher Childress
Like a real life "Indiana Jones," maverick archaeologist David Childress takes the reader on an incredible adventure across some of the world's oldest and most remote countries in search of lost cities and ancient mysteries. Discover ancient cities in the Gobi Desert; hear fantastic tales of lost continents, vanished civilizations and secret societies bent on ruling the world; visit forgotten monasteries in forbidding snow-capped mountains with strange tunnels to mysterious subterranean cities! A unique combination of far-out exploration and practical travel advice, it will astound and delight the experienced traveler or the armchair voyager.

429 PAGES. 6x9 PAPERBACK. ILLUSTRATED. FOOTNOTES & BIBLIOGRAPHY. $14.95. CODE: CHI

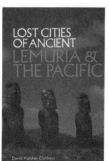

LOST CITIES OF ANCIENT LEMURIA & THE PACIFIC
by David Hatcher Childress
Was there once a continent in the Pacific? Called Lemuria or Pacifica by geologists, Mu or Pan by the mystics, there is now ample mythological, geological and archaeological evidence to "prove" that an advanced and ancient civilization once lived in the central Pacific. Maverick archaeologist and explorer David Hatcher Childress combs the Indian Ocean, Australia and the Pacific in search of the surprising truth about mankind's past. Contains photos of the underwater city on Pohnpei; explanations on how the statues were levitated around Easter Island in a clockwise vortex movement; tales of disappearing islands; Egyptians in Australia; and more.

379 PAGES. 6x9 PAPERBACK. ILLUSTRATED. FOOTNOTES & BIBLIOGRAPHY. $14.95. CODE: LEM

LOST CITIES & ANCIENT MYSTERIES OF THE SOUTHWEST
By David Hatcher Childress

Join David as he starts in northern Mexico and searches for the lost mines of the Aztecs. He continues north to west Texas, delving into the mysteries of Big Bend, including mysterious Phoenician tablets discovered there and the strange lights of Marfa. Then into New Mexico where he stumbles upon a hollow mountain with a billion dollars of gold bars hidden deep inside it! In Arizona he investigates tales of Egyptian catacombs in the Grand Canyon, cruises along the Devil's Highway, and tackles the century-old mystery of the Lost Dutchman mine. In Nevada and California Childress checks out the rumors of mummified giants and weird tunnels in Death Valley, plus he searches the Mohave Desert for the mysterious remains of ancient dwellers alongside lakes that dried up tens of thousands of years ago. It's a full-tilt blast down the back roads of the Southwest in search of the weird and wondrous mysteries of the past!

486 Pages. 6x9 Paperback. Illustrated. $19.95. Code: LCSW

TECHNOLOGY OF THE GODS
The Incredible Sciences of the Ancients
by David Hatcher Childress

Childress looks at the technology that was allegedly used in Atlantis and the theory that the Great Pyramid of Egypt was originally a gigantic power station. He examines tales of ancient flight and the technology that it involved; how the ancients used electricity; megalithic building techniques; the use of crystal lenses and the fire from the gods; evidence of various high tech weapons in the past, including atomic weapons; ancient metallurgy and heavy machinery; the role of modern inventors such as Nikola Tesla in bringing ancient technology back into modern use; impossible artifacts; and more.

356 PAGES. 6x9 PAPERBACK. ILLUSTRATED. BIBLIOGRAPHY. $16.95. CODE: TGOD

LOST CONTINENTS & THE HOLLOW EARTH
I Remember Lemuria and the Shaver Mystery
by David Hatcher Childress & Richard Shaver

Shaver's rare 1948 book *I Remember Lemuria* is reprinted in its entirety, and the book is packed with illustrations from Ray Palmer's *Amazing Stories* magazine of the 1940s. Palmer and Shaver told of tunnels running through the earth—tunnels inhabited by the Deros and Teros, humanoids from an ancient spacefaring race that had inhabited the earth, eventually going underground, hundreds of thousands of years ago. Childress discusses the famous hollow earth books and delves deep into whatever reality may be behind the stories of tunnels in the earth. Operation High Jump to Antarctica in 1947 and Admiral Byrd's bizarre statements, tunnel systems in South America and Tibet, the underground world of Agartha, the belief of UFOs coming from the South Pole, more.

344 PAGES. 6x9 PAPERBACK. ILLUSTRATED. $16.95. CODE: LCHE

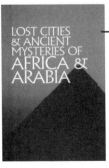

LOST CITIES & ANCIENT MYSTERIES OF AFRICA & ARABIA
by David Hatcher Childress

Childress continues his world-wide quest for lost cities and ancient mysteries. Join him as he discovers forbidden cities in the Empty Quarter of Arabia; "Atlantean" ruins in Egypt and the Kalahari desert; a mysterious, ancient empire in the Sahara; and more. This is the tale of an extraordinary life on the road: across war-torn countries, Childress searches for King Solomon's Mines, living dinosaurs, the Ark of the Covenant and the solutions to some of the fantastic mysteries of the past.

423 PAGES. 6x9 PAPERBACK. ILLUSTRATED. $14.95. CODE: AFA

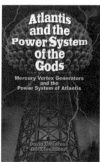

ATLANTIS & THE POWER SYSTEM OF THE GODS
by David Hatcher Childress and Bill Clendenon
Childress' fascinating analysis of Nikola Tesla's broadcast system in light of Edgar Cayce's "Terrible Crystal" and the obelisks of ancient Egypt and Ethiopia. Includes: Atlantis and its crystal power towers that broadcast energy; how these incredible power stations may still exist today; inventor Nikola Tesla's nearly identical system of power transmission; Mercury Proton Gyros and mercury vortex propulsion; more. Richly illustrated, and packed with evidence that Atlantis not only existed—it had a world-wide energy system more sophisticated than ours today.
246 PAGES. 6x9 PAPERBACK. ILLUSTRATED. $15.95. CODE: APSG

THE ANTI-GRAVITY HANDBOOK
edited by David Hatcher Childress

The new expanded compilation of material on Anti-Gravity, Free Energy, Flying Saucer Propulsion, UFOs, Suppressed Technology, NASA Cover-ups and more. Highly illustrated with patents, technical illustrations and photos. This revised and expanded edition has more material, including photos of Area 51, Nevada, the government's secret testing facility. This classic on weird science is back in a new format!
230 PAGES. 7x10 PAPERBACK. ILLUSTRATED. $16.95. CODE: AGH

ANTI-GRAVITY & THE WORLD GRID
Is the earth surrounded by an intricate electromagnetic grid network offering free energy? This compilation of material on ley lines and world power points contains chapters on the geography, mathematics, and light harmonics of the earth grid. Learn the purpose of ley lines and ancient megalithic structures located on the grid. Discover how the grid made the Philadelphia Experiment possible. Explore the Coral Castle and many other mysteries, including acoustic levitation, Tesla Shields and scalar wave weaponry. Browse through the section on anti-gravity patents, and research resources.
274 PAGES. 7x10 PAPERBACK. ILLUSTRATED. $14.95. CODE: AGW

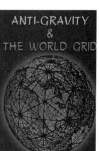
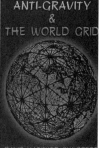

ANTI-GRAVITY & THE UNIFIED FIELD
edited by David Hatcher Childress
Is Einstein's Unified Field Theory the answer to all of our energy problems? Explored in this compilation of material is how gravity, electricity and magnetism manifest from a unified field around us. Why artificial gravity is possible; secrets of UFO propulsion; free energy; Nikola Tesla and anti-gravity airships of the 20s and 30s; flying saucers as superconducting whirls of plasma; anti-mass generators; vortex propulsion; suppressed technology; government cover-ups; gravitational pulse drive; spacecraft & more.
240 PAGES. 7x10 PAPERBACK. ILLUSTRATED. $14.95. CODE: AGU

THE TIME TRAVEL HANDBOOK
A Manual of Practical Teleportation & Time Travel
edited by David Hatcher Childress
The Time Travel Handbook takes the reader beyond the government experiments and deep into the uncharted territory of early time travellers such as Nikola Tesla and Guglielmo Marconi and their alleged time travel experiments, as well as the Wilson Brothers of EMI and their connection to the Philadelphia Experiment—the U.S. Navy's forays into invisibility, time travel, and teleportation. Childress looks into the claims of time travelling individuals, and investigates the unusual claim that the pyramids on Mars were built in the future and sent back in time. A highly visual, large format book, with patents, photos and schematics. Be the first on your block to build your own time travel device!
316 PAGES. 7x10 PAPERBACK. ILLUSTRATED. $16.95. CODE: TTH

MAPS OF THE ANCIENT SEA KINGS
Evidence of Advanced Civilization in the Ice Age
by Charles H. Hapgood
Charles Hapgood has found the evidence in the Piri Reis Map that shows Antarctica, the Hadji Ahmed map, the Oronteus Finaeus and other amazing maps. Hapgood concluded that these maps were made from more ancient maps from the various ancient archives around the world, now lost. Not only were these unknown people more advanced in mapmaking than any people prior to the 18th century, it appears they mapped all the continents. The Americas were mapped thousands of years before Columbus. Antarctica was mapped when its coasts were free of ice!
316 PAGES. 7x10 PAPERBACK. ILLUSTRATED. BIBLIOGRAPHY & INDEX. $19.95. CODE: MASK

PATH OF THE POLE
Cataclysmic Pole Shift Geology
by Charles H. Hapgood
Maps of the Ancient Sea Kings author Hapgood's classic book *Path of the Pole* is back in print! Hapgood researched Antarctica, ancient maps and the geological record to conclude that the Earth's crust has slipped on the inner core many times in the past, changing the position of the pole. *Path of the Pole* discusses the various "pole shifts" in Earth's past, giving evidence for each one, and moves on to possible future pole shifts.
356 PAGES. 6x9 PAPERBACK. ILLUSTRATED. $16.95. CODE: POP

SECRETS OF THE HOLY LANCE
The Spear of Destiny in History & Legend
by Jerry E. Smith
Secrets of the Holy Lance traces the Spear from its possession by Constantine, Rome's first Christian Caesar, to Charlemagne's claim that with it he ruled the Holy Roman Empire by Divine Right, and on through two thousand years of kings and emperors, until it came within Hitler's grasp—and beyond! Did it rest for a while in Antarctic ice? Is it now hidden in Europe, awaiting the next person to claim its awesome power? Neither debunking nor worshiping, *Secrets of the Holy Lance* seeks to pierce the veil of myth and mystery around the Spear. Mere belief that it was infused with magic by virtue of its shedding the Savior's blood has made men kings. But what if it's more? What are "the powers it serves"?
312 PAGES. 6x9 PAPERBACK. ILLUSTRATED. BIBLIOGRAPHY. $16.95. CODE: SOHL

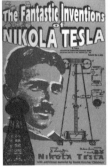

THE FANTASTIC INVENTIONS OF NIKOLA TESLA
by Nikola Tesla with additional material by David Hatcher Childress
This book is a readable compendium of patents, diagrams, photos and explanations of the many incredible inventions of the originator of the modern era of electrification. In Tesla's own words are such topics as wireless transmission of power, death rays, and radio-controlled airships. In addition, rare material on a secret city built at a remote jungle site in South America by one of Tesla's students, Guglielmo Marconi. Marconi's secret group claims to have built flying saucers in the 1940s and to have gone to Mars in the early 1950s! Incredible photos of these Tesla craft are included. •His plan to transmit free electricity into the atmosphere. •How electrical devices would work using only small antennas. •Why unlimited power could be utilized anywhere on earth. •How radio and radar technology can be used as death-ray weapons in Star Wars.
342 PAGES. 6x9 PAPERBACK. ILLUSTRATED. $16.95. CODE: FINT

ORDER FORM

10% Discount When You Order 3 or More Items!

One Adventure Place
P.O. Box 74
Kempton, Illinois 60946
United States of America
Tel.: 815-253-6390 • Fax: 815-253-6300
Email: auphq@frontiernet.net
http://www.adventuresunlimitedpress.com

ORDERING INSTRUCTIONS

✓ Remit by USD$ Check, Money Order or Credit Card

✓ Visa, Master Card, Discover & AmEx Accepted

✓ Paypal Payments Can Be Made To:

info@wexclub.com

✓ Prices May Change Without Notice

✓ 10% Discount for 3 or More Items

SHIPPING CHARGES

United States

✓ Postal Book Rate { $4.50 First Item / 50¢ Each Additional Item

✓ POSTAL BOOK RATE Cannot Be Tracked!
Not responsible for non-delivery.

✓ Priority Mail { $6.00 First Item / $2.00 Each Additional Item

✓ UPS { $7.00 First Item / $1.50 Each Additional Item

NOTE: UPS Delivery Available to Mainland USA Only

Canada

✓ Postal Air Mail { $15.00 First Item / $3.00 Each Additional Item

✓ Personal Checks or Bank Drafts MUST BE

US$ and Drawn on a US Bank

✓ Canadian Postal Money Orders OK

✓ Payment MUST BE US$

All Other Countries

✓ Sorry, No Surface Delivery!

✓ Postal Air Mail { $19.00 First Item / $7.00 Each Additional Item

✓ Checks and Money Orders MUST BE US$
and Drawn on a US Bank or branch.

✓ Paypal Payments Can Be Made in US$ To:
info@wexclub.com

SPECIAL NOTES

✓ RETAILERS: Standard Discounts Available

✓ BACKORDERS: We Backorder all Out-of-
Stock Items Unless Otherwise Requested

✓ PRO FORMA INVOICES: Available on Request

✓ DVD Return Policy: Replace defective DVDs only

ORDER ONLINE AT: www.adventuresunlimitedpress.com

10% Discount When You Order 3 or More Items!

Please check: ☑

| ☐ This is my first order | ☐ I have ordered before |

Name

Address

City

State/Province | Postal Code

Country

Phone: Day | Evening

Fax | Email

Item Code	Item Description	Qty	Total

Please check: ☑

	Subtotal ▶	
	Less Discount-10% for 3 or more items ▶	
☐ Postal-Surface	Balance ▶	
☐ Postal-Air Mail (Priority in USA)	Illinois Residents 6.25% Sales Tax ▶	
	Previous Credit ▶	
☐ UPS	Shipping ▶	
(Mainland USA only)	Total (check/MO in USD$ only) ▶	

☐ Visa/MasterCard/Discover/American Express

Card Number:

Expiration Date: | Security Code:

✓ SEND A CATALOG TO A FRIEND: